T0275724

Integrated Management of Salt Affected Soils in Agriculture

Integrated Management of Salt Affected Soils in Agriculture

Incorporation of Soil Salinity Control Methods

Nesreen Abou-Baker
National Research Center (NRC),
Agricultural and Biological Division,
Soils and Water Use Dept.

Ebtisam El-Dardiry
National Research Center (NRC),
Agricultural and Biological Division,
Water Relations and Irrigation Dept.

AMSTERDAM • BOSTON • HEIDELBERG • LONDON
NEW YORK • OXFORD • PARIS • SAN DIEGO
SAN FRANCISCO • SINGAPORE • SYDNEY • TOKYO

Academic Press is an imprint of Elsevier

Publisher: Nikki Levy
Senior Acquisition Editor: Nancy Maragioglio
Editorial Project Manager: Billie Jean Fernandez
Production Manager: Anusha Sambamoorthy

Academic Press is an imprint of Elsevier
125, London Wall, EC2Y 5AS.
525 B Street, Suite 1800, San Diego, CA 92101-4495, USA
225 Wyman Street, Waltham, MA 02451, USA
The Boulevard, Langford Lane, Kidlington, Oxford OX5 1GB, UK

ISBN: 978-0-12-804165-9

Library of Congress Cataloging-in-Publication Data
A catalog record for this book is available from the Library of Congress

British Library Cataloguing-in-Publication Data
A catalogue record for this book is available from the British Library

For Information on all Academic Press publications
visit our website at http://store.elsevier.com/

Working together
to grow libraries in
developing countries

www.elsevier.com • www.bookaid.org

CONTENTS

ACKNOWLEDGEMENTS

Appreciation is expressed to all the scientists who are listed in the references sections of this book for their efforts in the expansion of knowledge.

Environmental stresses, such as salinity and drought, cause huge losses to agricultural production which can consequently affect the economic strengths of countries. Soil salinity is the most widespread abiotic stress in the arid and semiarid regions. It is responsible for osmotic imbalance, declining water availability for plants, causing ion-specific toxicities or imbalance, decreasing nutrients uptake, and thus, affecting plant physiology (photosynthesis activity, inhibiting plants' metabolisms and chlorophyll content). The integrated management of salt-affected soils relates to controlling the problems of soil salinity, amelioration, remediation, rehabilitation, or reclaiming salt-affected soils by the collection of some effective methods of soil salinity control.

The mission of this review book is to develop the proper management procedures and to solve the problems of crop production on salt-affected soils through research, new knowledge, and technology. I hope this short book will be useful to those farmers who are suffered from salinity stress and will encourage both young scientists and researchers to continue studying soil salinity control and trying to find the most suitable combination of the different methods to develop an integrated management of salt affected soils. The book may encourage government agencies (Such as Ministry of Agriculture) to guiding the farmers who suffering from soil salinity problems by issuing guidance leaflets containing integrated recommendation for salt affected soils reclamation.

Nesreen Abou-Baker

INTRODUCTION

Arid and semiarid regions of the world are generally associated with high population density and lower than average per capita incomes and living standards. These regions are vulnerable to food shortages due to the current, unsustainable use of land affected by soil salinization. It is worth mentioning that most of the new reclaimed areas in Egypt are salinity affected (Dewdar and Rady, 2013).

Environmental stress conditions such as drought, heat, salinity, cold, or pathogen infection can have a devastating impact on plant growth and yield under field conditions (Suzuki et al., 2014). Abiotic stresses cause considerable loss to agricultural production worldwide (Shao et al., 2008). Soil salinity is most widespread abiotic stress in the arid and semiarid regions and also a serious problem in areas where groundwater of high salt content is used for irrigation. The most serious salinity problems are being faced in the irrigated arid and semi-arid regions of the world and it is in these very regions that irrigation is essential to increase agricultural production to satisfy food requirements.

Salinity is responsible for osmotic imbalance, declining soil water availability for plants, causing ion-specific toxicities or imbalance, decreasing nutrients uptake, and thus, affecting plant physiology (photosynthesis activity, inhibiting plants' metabolisms, and chlorophyll content). All of these changes contribute to stunted growth and reduced productivity of plants, and consequently, the weakness of the economy of countries suffering from salinity. Using the FAO/UNESCO soil map of the world (1970–1980), FAO estimated that globally the total area of saline soils was 397 million ha and that of sodic soils 434 million ha. Of the then 230 million ha of irrigated land, 45 million ha (19.5%) were salt-affected soils; and of the almost 1500 million ha of dryland agriculture, 32 million (2.1%) were salt-affected soils (FAO, 2015). Worldwide, more than 800 million hectares of soils are salt-affected (FAO, 2008; Rengasamy, 2010).

Soil salinity control relates to controlling the problems of soil salinity and reclaiming salinized agricultural land. The aim of soil salinity control is to prevent soil degradation by salinization and reclaim already salty (saline) soils. Soil reclamation is also called soil improvement, rehabilitation, remediation, recuperation, or amelioration (Rengasamy, 2010).

The mission of this review book is to develop the proper management procedures and to solve problems of crop production on salt-affected soils through research, new knowledge, and technology.

CHAPTER *1*

Salt-Affected Soil Definition and Types

1.1 SALT-AFFECTED SOILS DEFINITION

Different types of physical, chemical, and/or biological land degradation processes (e.g., compaction, inorganic/organic contamination, diminished microbial activity/diversity), mostly under excessive anthropogenic pressures during the last century, have resulted in serious consequences to global natural resources. Among them, soil salinization, arising from either natural or human-induced causes, led to an increase in concentration of dissolved salts in the soil profile to a level that impairs food production, environmental health, and socioeconomic well-being (Ondrasek et al., 2011).

Wind-transported materials from soil or lake surfaces are another source of salts. Application of fertilizers and soil amendments, poor quality irrigation water and capillary rise of shallow saline groundwater, insufficient water application, and insufficient drainage can all contribute to the salinization of the soil layers. Even seawater intrusion onto land, which is a growing problem as sea levels rise in many parts of the world, can deposit a large amount of salts in soils of coastal lands (FAO, 2005; Rengasamy, 2010).

1.2 TYPES OF SALT-AFFECTED SOILS

The categories are defined in soil classification literature by the Natural Resources Conservation Service (NRCS) (Table 1.1). EC, electrical conductivity; SAR, sodium adsorption ratio; ESP, exchangeable sodium percentage. Three general categories of salt-affected soils have been identified for management purposes (FAO, 2005, 2008; Horneck et al., 2007; Waskom et al., 2014). Figure 1.1 illustrates a salinized patch in a wheat crop.

Integrated Management of Salt Affected Soils in Agriculture. DOI: http://dx.doi.org/10.1016/B978-0-12-804165-9.00001-7

Table 1.1 Classification of Salt-Affected Soils

Soil Properties	Non	Saline	Sodic	Saline – Sodic
EC dS/m	Below 4	Above 4	Below 4	Above 4
SAR	Below 13	Below 13	Above 13	Above 13
ESP	Below 15	Below 15	Above 15	Above 15
Soil structure	Flocculated	Flocculated	Dispersed	Flocculated
Soil physical condition	Normal	Normal	Poor	Normal
Symptoms	No adverse effect on soil. Healthy plants	White crust on soil surface. Water-stressed plants. Leaf tip burn.	Poor drainage. Black powdery residue on soil surface.	Grey-colored soil. Plants showing water stress.

EC: electrical conductivity, SAR: sodium adsorption ratio, ESP: exchangeable sodium percentage.

Figure 1.1 A salinized patch in a wheat crop.

- Saline soils: Salt problems in general
- Sodic soils: Sodium problems
- Saline–sodic soils: Problems with sodium and other salts.

CHAPTER 2

Features, Nature, and Distribution of Salt-Affected Soils

2.1 FEATURES OF SALT-AFFECTED SOILS

MacCauley and Jones (2005) summarized the features that appear on salt-affected soils in the following:

- *Loss of productive land*: As the salt concentration increases, the energy of water in the soil decreases such that it is more and more difficult for plants to extract that water, even though the soil seems relatively moist, and the plants are unable to take up the water and nutrients they require. Hence, soils with too much soluble salt in the root zone effectively create a "chemical induced drought" and the plants may wilt and die.
- *Nutrient supply*: Na^+ and Cl^- ions can be toxic. Some plants (e.g., barley) are able to restrict the entry of these ions, but others (e.g., chickpea) cannot. Localized salinity damage can temporarily occur through the application of soluble fertilizers too close to salt-sensitive seedlings under dry conditions.
- *Soil structure decline*: Salinity is often associated with areas of structural decline and increased risk of erosion. Sodic soils are commonly found in association with NaCl.
- *Susceptibility to erosion*: Salinity reduces ground cover, often leaving soil bare and can contribute to an increased risk of wind and water erosion depending on soil texture.
- *Reduced water supplies*: The mobilization of salt through changes in land use, can lead to a reduction in the quality of both surface water and groundwater supplies affecting agricultural production and environmental quality.
- *Loss of biodiversity*: Increasing salinity causes death of native vegetation, declines in wetland health, shifts in the ecological balance of species or loss of ecosystem function, and a loss of biological diversity.

Integrated Management of Salt Affected Soils in Agriculture. DOI: http://dx.doi.org/10.1016/B978-0-12-804165-9.00002-9

2.2 NATURE AND DISTRIBUTION OF SALT-AFFECTED SOILS

Salt-affected soils occur in more than 100 countries of the world with a variety of extents, nature, and properties (Rengasamy, 2005). According to Tòth et al. (2008), the total area of salt-affected soil is about 1 billion hectares. They occur mainly in the arid and semiarid regions of Asia, Australia, Africa, and South America. Nowadays, 10—15% of the irrigated land (45 Mha) suffers with different levels of salinization problems and 0.5—1% of irrigated area is lost every year (FAO, 1996). Table 2.1 shows the potential risks of salinization all over the world (Lakhdar et al., 2009).

The majority of salt-affected soils in Egypt are located in the Northern-Central part of the Nile Delta and on its Eastern and Western sides. Other areas are found in Wadi El-Natroun, Tal El-Kebeir and the Oases, many parts of the Nile Delta and Valley, and El-Fayoum province. Nine hundred thousand hectares suffer from salinization problems in cultivated irrigated areas. Sixty percent of the cultivated lands of the Northern Delta region are salt-affected, twenty percent of the Southern Delta and Middle Egyptian region, and twenty five percent of the Upper Egypt region are salt-affected soils (Gehad, 2003).

Soils of primary salinity are mainly those of marine and lacustrine origin, such as brackish lake beds or salt marshes along the northern coast (soils of Edko, Maruit, and Burullos). These soils are saline or

Table 2.1 The Potential Risks of Salinization in all Over the World (Lakhdar et al., 2009)		
Country	Salt-Affected Area	References
Australia	30% total area	Rengasamy (2005)
Egypt	9.1% total area	Mashali et al. (2005)
Hungary	10% total area	Varallyay (1992)
Iran	28% irrigated land	Khel (2006)
Kenya	14.4% total area	Mashali et al. (2005)
Nigeria	20% irrigated land	FAO (2000)
Russia	21% agricultural land	Dobrovol'skii and Stasyuk (2008)
Syria	40% irrigated land	FAO (2000)
Thailand	30% total area	Yuvaniyama (2001)
Tunisia	11.6% total area	Mashali et al. (2005)
USA	25—30% irrigated land	Wichelns (1999)

saline-alkali with sodium chloride and sulfate as dominant salts, while in Fayoum, the salinization took place as a result of the deposition of migrating salts during the soil formation. Also, two other types of deteriorated soils were identified: Saline and saline-alkali soils which are characterized by their adverse physical properties, their dispersed structure and impermeability to water, are to be directly connected with sodium as the dominant exchangeable base and the presence of magnesium silicate precipitated during the process of soil alkalinization.

CHAPTER 3

Soil Salinization and Sodification Factors

3.1 NATURAL FACTORS

Natural, or "primary salinity" occurs in arid climates throughout the world. These areas include salt marshes, salt lakes, and salt flats, all of which are considered ecologically significant.

1. *Parent material*: Many geological formations beneath soils used for agriculture have marine origins and often contain high levels of salts (NaCl). Geological phenomena may increase the salts concentration in groundwater and consequently in the soil.
2. *Earthworks*: Causes of salinity include barriers to the movement of water which can include roads, railways, and land reformation.
3. *Hydrological processes*: Fluctuations in depth to a watertable are influenced by the amount of recharge (water entry) and discharge (water loss). Hillel (2004) stated that evaporation at the surface leaves behind concentrated salt deposits known as "saline scalds."
   a. Natural factors capable of bringing groundwater containing elevated salt contents to the surface.
   b. Infiltration of groundwater in below sea level zones (microdepressions with reduced or absent drainage).
   c. Drainage of waters from zones with geological substrates capable of liberating large amounts of salts.
4. *Climate*: In many regions of arid and semiarid regions, potential evaporation is higher than the amount of rainfall. Consequently, most of the water that infiltrates into soils is extracted by plants to be transpired or it evaporates from the soil surface leaving its salt content in the soil. In addition to the action of winds, this, in coastal zones, can transport moderate amounts of salts to the interior. Under salt-affected soils, salinization is a part of the natural ecosystem in arid and semiarid regions and an increasing problem in agricultural soils the world over. In temperate, moist climates salinity occurs on a smaller scale, principally in salt water marshes

Integrated Management of Salt Affected Soils in Agriculture. DOI: http://dx.doi.org/10.1016/B978-0-12-804165-9.00003-0

(Qadir et al., 2006). Also, Blanco and Lal (2010) reported that salts originate from the natural weathering of minerals or from fossil salt deposits left from ancient sea beds. Salts accumulate in the soil of arid climates as irrigation water or groundwater seepage evaporates, leaving minerals behind.

5. *Over-clearing of deep-rooted vegetation*: Replacement of deep-rooted native vegetation with shallower rooted annual crops results in less evapotranspiration (ET_0) over the whole year. Consequently, there is greater percolation of water below the root zone. Droughts greatly reduce deep drainage losses from agricultural systems, causing the expansion of salt-affected land to be slowed.

3.2 ANTHROPOGENIC FACTORS

1. *Irrigation*: Water storages and delivery channels associated with irrigation developments often leak and cause the local watertable to rise and surface salinity to develop. However, most of the salinization under irrigation comes from inefficient irrigation practices whereby water much in excess of a crop's needs is applied over a number of years. Irrigation with water containing elevated salt content raises salinity. The excessive amounts of salts provided by irrigation water can have adverse effects on the chemical and physical properties of the soils and on their biological processes (Tejada et al., 2006). Rhoades and Loveday (1990) added that thus salinization of a soil depends on the quality of the irrigation water used (Table 3.1), on the existence and level of natural and/or artificial drainage of the soil, on the depth of the

Table 3.1 Approximate Water Salinity Classes		
Water Classes	EC (dS/m)	Total Dissolved Salts (g/L)
Nonsaline	<0.7	<0.5
Slightly saline	0.7–2.0	0.5–1.5
Moderately saline	2.0–10.0	1.5–7.0
Highly saline	10.0–20.0	7.0–15.0
Very highly saline	20.0–45.0	15.0–35.0
Brine	>45.0	>35.0
Source: *Rhoades and Loveday (1990)*.		

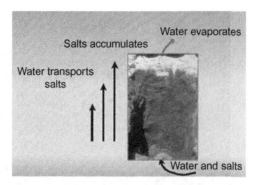

Figure 3.1 Salt crust formation.

water-bearing stratum and on the original concentration of salt in the soil profile.

Soils affected by salts commonly appear in irrigated areas due to inadequate management of the irrigation and other practices, such that important extensions of fertile, arable land are becoming more and more saline (FAO, 2008).

2. Rise in phreatic water level due to human activities (infiltration of water from unlined channels and reservoirs, irregular distribution of irrigation water, deficient irrigation practices, and inadequate drainage). Figure 3.1 illustrates salt crust formation.

3. *Fertilization*: Fertilizers used and other production factors, namely for intensive agriculture in land with low permeability and reduced possibilities for leaching, also the incorrect and excessive use of chemical fertilizers. Moreover, mineralization of the carbon and nitrogen and the enzymatic activity, which is crucial for the decomposition of organic matter and liberation of the nutrients necessary for sustainability of the production (Azam and Ifzal, 2006). In addition, the agricultural practices can increase or reduce the microbial population, thus altering the activity, source, and persistence of the enzymes in the soil.

4. Contamination of the soil with industrial water and subproducts with high salt contents.

The factors which can affect on salinization area formation and distribution are illustrated in Figure 3.2.

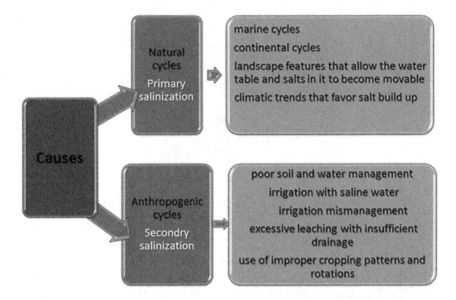

Figure 3.2 Causes of salt affected soils formation.

CHAPTER 4

Effect of Salinity on Soil Microorganisms, Plant Growth, and Yield

4.1 EFFECT OF SALINITY ON SOIL MICROORGANISMS

The microbial communities of the soil perform a fundamental role in cycling nutrients, in the volume of organic matter in the soil, and in maintaining plant productivity. Stress can be detrimental for sensitive microorganisms and decrease the activity of surviving cells, due to the metabolic load imposed by the need for stress tolerance mechanisms (Yuan et al., 2007). In a dry hot climate, the low humidity and soil salinity are the most stressful factors for the soil microbial flora, and frequently occur simultaneously. Saline stress can gain importance, especially in agricultural soils where the high salinity may be a result of irrigation practices and the application of chemical fertilizers.

Two strategies are used by microorganisms to adapt to osmotic stress, both of which result in an accumulation of solutes in the cell to counteract the increase in osmotic pressure (Tripathi et al., 2006). One is the selective exclusion of the solute incorporated (e.g., Na^+, Cl^-), thus accumulating the ions necessary for metabolism (e.g., NH_4^+). High salinity reduces the microbial biomass and the availability of nutrients for plants is regulated by the rhizospheric microbial activity. Thus any factor affecting this community and its functions influences the availability of nutrients and growth of the plants. Hussein and Abou-Baker (2014) concluded that significant depressions were obtained in nitrogen concentration or content as a result of growing moringa plants under salinity conditions. There was a similar response in phosphorus content but the differences were not significant. Calcium and potassium concentrations did not significantly responded with salinity but magnesium concentration decreased significantly only with the first level of salinity.

Integrated Management of Salt Affected Soils in Agriculture. DOI: http://dx.doi.org/10.1016/B978-0-12-804165-9.00004-2

4.2 EFFECT OF SALINITY ON PLANT GROWTH AND YIELD

Excessive soil salinity reduces the yield of many crops. This ranges from a slight crop loss to complete crop failure, depending on the type of crop and the severity of the salinity problem. Although several treatments and management practices can reduce salt levels in the soil, there are some situations where it is either impossible or too costly to attain desirably low soil salinity levels (Qadir and Oster, 2004). In some cases, the only viable management option is to plant salt-tolerant crops. Sensitive crops, such as pinto beans, cannot be managed profitably in saline soils. Cardon et al. (2014) reported that fruit crops may show greater yield variation because a large number of rootstocks and varieties are available. Also, the stage of plant growth has a bearing on salt tolerance. Plants are usually most sensitive to salt during the emergence and early seedling stages (Davis et al., 2012). Tolerance usually increases as the crop develops. Table 4.1 shows the degree of soil salinity, expected land use, plant response and species.

The adverse effects of salinity on plant growth are the result of changes in plant physiology which include ion toxicity, osmotic stress, and nutrient deficiency (Frary et al., 2010). However, Roshandel and Flowers (2009) have shown that the changes induced by salinity are ionic rather than osmotic. Figure 4.1 illustrates how does soil salinity affect plant growth?

Salt stress, like other abiotic stresses, can also lead to oxidative stress due to increased production of reactive oxygen species (ROS), such as singlet oxygen ($^1O_2$), superoxide anion ($^.O_2^-$), hydrogen peroxide (H_2O_2), and hydroxyl radical ($^.OH$) (Bray et al., 2000). These ROS are highly reactive and can alter normal cellular metabolism with oxidative damage to carbohydrates, proteins, and nucleic acids, and cause peroxidation of membrane lipids (Azevedo Neto et al., 2008). The deleterious effects of salinity on plant growth are associated with low osmotic potential of soil solution causing physiological drought, nutritional imbalances, and specific ion toxicity or a combination of all these factors (Batool et al., 2014).

The cropping pattern in Egypt is somewhat adjusted to soil condition. In the Northern Delta where soil salinity is somewhat higher than

Table 4.1 Degree of Soil Salinity, Expected Land Use, Plant Response and Species

Degree of Salinity	ECe (dSm^{-1})	Level of Effect	Site Characteristics	Use	Plant Response	Species
Low salinity	2 to 4	Salinity effects usually minimal	Natural salinity; often seasonally dry irrigation salinity; can be waterlogged after irrigation	Cropping	Very little effect on all plants	Low-moderate salt tolerance
Moderate salinity	4–8	Yield of salt sensitive plants restricted	Dryland salinity; often seasonally waterlogged	Crop-pasture rotations	Some effects on salt sensitive crops	Moderate-salt tolerance
High salinity	8–16	Only salt tolerant plants yield satisfactory	Discharge areas; can be seeping or dry according to season	Grazing or re-vegetation	Considerable effects on salt sensitive crops	High salt tolerance
Very high salinity	<16	Few salt tolerant plants yield satisfactory	White crust on soil surface	Very few plants will tolerate and grow	Some effects on salt tolerant crops	Halophytes

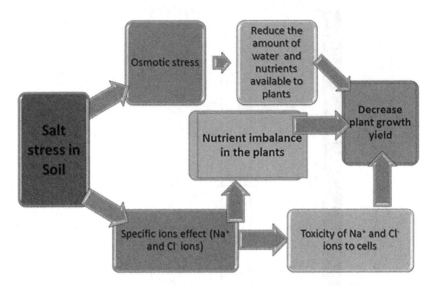

Figure 4.1 How does soil salinity affect plant growth?

normal, crop rotation includes rice and cotton as the main summer crops and wheat and clover as the main winter crops. All these crops have proven to be tolerant or semi-tolerant to salinity. Sugar beet which is known to be tolerant to salinity has been recently grown in the Northern Delta.

CHAPTER 5

Management and Rehabilitation of Salt-Affected Soils

5.1 SCRAPING AND REMOVAL OF SURFACE SOIL

Due to continuous evaporation the salt concentration is the highest in the surface soil. The top soil can be scraped and transported out of the field (FAO, 2005). Land leveling makes it possible for a more uniform application of water for better leaching and salinity control. For coarse leveling, simple scrapers of levelers may be used, while for fine leveling, the use of laser guidance has recently been adopted which is more effective (precision 1–3 cm), and restores or improves water infiltration (Gehad, 2003). Figure 5.1 illustrates the scraping of surface soil.

5.2 REPLACEMENT OF SODIUM IN THE SOIL

Sodium (Na^+), as the most frequent causative agent of salinity, is the most pronounced destructor (by dispersion) of secondary clay minerals. Dispersion occurs because of Na^+ replacement of calcium (Ca^{2+}) and other coagulators (Mg^{2+}, OM) adsorbed on the surface and/or interlayers of soil aggregates. The replacement is mainly caused by specific physical characteristics, such as the electrical charge and hydration ability of particular elements, whereas some chemical properties (e.g., increased pH) may facilitate the clay dispersion process (Ondrasek et al., 2011). Calcium has an effective role of replacing Na^+ in a saline rhizosphere. Significant positive influences of applied calcium paste were observed on growth traits (number of leaves, leaf area, and leaf dry weight per plant), chemical analysis (photosynthetic pigments, photosynthates, and nutrients) and seed cotton yield when compared with the control (without calcium paste) as described by Dewdar and Rady (2013). Figure 5.2 illustrates that replacing sodium with calcium in soil.

Replacement of exchangeable sodium ions can be accomplished by increasing the calcium concentration in the soil solution by the

Integrated Management of Salt Affected Soils in Agriculture. DOI: http://dx.doi.org/10.1016/B978-0-12-804165-9.00005-4

Figure 5.1 Scraping of surface soil.

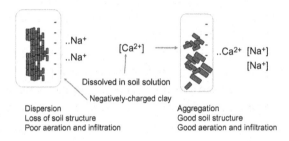

Figure 5.2 Replacement of sodium in soil with calcium.

addition of a calcium amendment (Davis et al., 2012). The ions on the soil particles are in a dynamic equilibrium with the soil solution. Therefore, a high concentration of calcium in the soil solution will result in calcium replacing sodium on the clay surfaces, and in the process, lowering the sodium adsorption ratio (SAR). Alberta Environment (2000) added that after a recent spill, the ionic concentration (EC) of the soil solution should not be allowed to fall below the level at which dispersion occurs until sufficient calcium has been applied to replace most of the sodium on the cation exchange complex and lower the SAR to acceptable levels. Ongoing monitoring of EC and SAR indicate the progress of the remediation process.

Calcium in its carbonate form is insoluble in water. Carbonic acid (H_2CO_3) is formed by the reaction of carbon dioxide (CO_2) and water (H_2O) and is produced by the plant roots' respiration and the decomposition of organic matter. Calcium in the presence of carbonic acid forms lime.

$$Ca + H_2CO_3 \rightarrow CaCO_3 \text{ (lime)}$$

Calcium in its sulfate form is soluble in water; however, calcium sulfate (gypsum) can precipitate at high rates. Gypsum ($CaSO_4 \cdot 2H_2O$) is used by many to treat a lime ($CaCO_3$) problem in soil. Sahin et al. (2011) reported that the time and effort that are required for in situ soil remediation will vary greatly depending on the site and spill conditions. Reclamation procedures vary with the characteristics of the spill (volume, severity, age, extent, etc.) and the site (soil type, topography, groundwater characteristics, soil permeability, selected plants, etc.). It may take several years and several treatments to return an area to productive use. Factors that will increase the time and effort required include: (i) impermeable soils (including dispersed soil, hard pans compacted layer, etc.), (ii) fine textured soils, (iii) shallow groundwater table, (iv) high EC/SAR levels, (v) low precipitation, (vi) pathways which promote movement to receptors (such as external drainage), (vii) remediation effort (e.g., number of amendment applications), and (viii) quality of leaching water.

Calcium amendments can be added to the soil in dry or liquid form. Liquid calcium amendments are faster acting, and have a deeper initial penetration depth. Commercial formulations of liquid calcium amendments are available in concentrated form. The most commonly used dry amendments are gypsum ($CaSO_4 \cdot 2H_2O$) and calcium nitrate, although calcium chloride may be used if adequate drainage control is in place and leachate is collected for proper disposal. Gypsum, which is moderately soluble, readily available, and inexpensive, is the most commonly applied calcium amendment. In addition to supplying calcium for the replacement of exchangeable sodium, the sustained release of electrolyte from gypsum contributes to the maintenance of the hydraulic conductivity (HC) and assists remediation of saline—sodic soil (Davis et al., 2012). Figure 5.2 illustrates the adverse effect of sodium on soil.

Calcium nitrate or calcium chloride minerals can be used to reclaim sodic soils, but they generally are more costly and are likely to produce other negative effects on plant growth or the environment. Nitrate is considered a groundwater contaminant and is not a good choice. Zaka et al. (2005) mentioned that limestone is another commonly available mineral that contains calcium. However, it is not used for reclaiming sodic soils because it is not soluble at the high pH levels common in these soils.

Elemental sulfur (S) can be used for sodic soil reclamation. Sulfur should be used only if free lime already exists in the soil. The addition of sulfur does not directly add calcium to the soil. However, sulfur will be oxidized biologically to form sulfuric acid, which dissolves lime (calcium carbonate, $CaCO_3$), often existing in arid and semiarid zone soils. The dissolution of indigenous lime provides the calcium necessary to reclaim a sodic soil. When adequate moisture and temperatures are present, oxidation of elemental sulfur will be completed within one or two growing seasons.

$S + $ *Thiobacillus* $ + $ warm and wet soil $\rightarrow H_2SO_4$

$H_2SO_4 + CaCO_3$ (lime) $\rightarrow CaSO_4 \cdot 2H_2O$

Sulfur is an important plant nutrient involved in plant growth and development. It is considered fourth in importance after nitrogen, phosphorus, and potassium. It is an integral part of several important compounds, such as vitamins, coenzymes, phytohormones, and reduced sulfur compounds that decipher growth and vigor of plants under optimal and stress conditions (Nazar et al., 2011). Adequate sulfur nutrition improves photosynthesis and growth of plants, and it has a regulatory interaction with nitrogen assimilation (Scherer, 2008). It is required for protein synthesis and, being a structural constituent of several coenzymes and prosthetic groups, it is also significant for nitrogen assimilation (Marschner, 1995). The assimilatory pathways of sulfur and nitrogen have been considered functionally convergent and well-coordinated as the availability of one element regulates the other (Schnug et al., 1993).

Elemental sulfur as a soil amendment should be well incorporated to increase the rate of reaction and to speed reclamation. When elemental sulfur is left on the soil surface, or when the soil is dry or cold, microbial conversion of elemental sulfur to sulfuric acid is delayed. Abdelhamid et al. (2013) found that most of the studied soil hydrophysical properties were improved with sulfur application in addition to plant growth. Carrow and Duncan (2011) showed the gypsum and sulfur element rates which needed to reclaim of Sodic Soils as presented in Table 5.1.

Abd El-Hady and Shaaban (2010) used another technique that improves salt-affected soil and eliminates salinity affect and improves hydrophysical properties, such as HC, through acidification of irrigation water.

Table 5.1 Gypsum and Sulfur Element Rates Which Needed to Reclaim of Sodic Soils (Carrow and Duncan, 2011)				
Exchangeable Na to be Replaced by Ca (meq Na/100 g soil)	Gypsum (ton/fed)		Elemental Sulfur (ton/fed)	
	300 mm	150 mm	300 mm	150 mm
1	1.8	0.9	0.32	0.16
2	3.4	1.7	0.64	0.32
4	6.9	3.4	1.28	0.64
8	13.7	6.9	2.56	1.28

On the first application of a calcium amendment, it is often beneficial to apply both faster acting calcium nitrate and slower release gypsum together. The amount of calcium nitrate applied is often limited by concerns about cost and nitrate contamination of groundwater. El-Hady and Abo-Sedera (2006a) found that addition of gypsum increased the HC from 0.00 to 0.06 mm/h in the soil with an exchangeable sodium percentage (ESP) of 32. Also when polymers were used in conjunction with gypsum, HC increased to 0.28 mm/h. At ESP values >15, an additional mechanism that may have been controlling HC was swelling, and none of the polymers reduced soil swelling.

Davis et al. (2012) summarized types of problems and the suitable action in reclamation of salt-affected soils. When only some of the reclamation problems are considered, amending soils or leaching can worsen the situation. When considering the whole system drainage, pH, salts, and sodium are the main concerns.

The principal objective of the recovery of soils affected by salts is to reduce the concentration of soluble salts and of exchangeable sodium in the soil profile, to a level that does not prejudice the development of crops. A decrease in the degree of salinity involves the process of dissolution and consequent removal by percolation water, whereas a decrease in the exchangeable sodium content involves its displacement from the exchange complex by calcium before the leaching process. Gharaibeh et al. (2009) added the application of gypsum by 2% and increased potato yield under 0.6% salt by more than 1.2%. The substituted sodium is leached from the radicular zone by way of excess irrigation, a process that demands an adequate flow of water through the soil (Qadir et al., 2006).

The low solubility of Ca^{2+} during remediation could limit its efficiency, and thus the possibility of using it with microorganisms is being explored so as to provide more active Ca^{2+} from plaster. Experiments carried out with blue-green algae and plaster resulted in greater solubility of the plaster, thus providing recovery of the sodic soils. Also, Sahin et al. (2011) mentioned that the measurement of the saturated HC of the soil columns after treatment by plaster indicated that it increased significantly in the saline−sodic soils after application of the microorganisms. The data suggest that the microorganisms tested could have the potential to help improve water circulation through saline soils.

5.3 LEACHING PLUS ARTIFICIAL DRAINAGE

The direct aim of the reshaping of field drains is to lower the watertable level, and optimize soil physical environment, and consequently, the removal of salts (including Na) in the soil solution by leaching with natural precipitation or irrigation. This step may involve collection and proper disposal of leachate, and must be carried out in the above order to avoid dispersion and further deterioration of soil conditions, as well as to facilitate remediation (Davis et al., 2012). Figure 5.3 illustrates the reshaping of field drains.

There are three ways to manage saline soils as described by Noory et al. (2009) as follows: (i) salts can be moved below the root zone by applying more water than the plant needs (leaching requirement (LR) method); (ii) where soil moisture conditions dictate, combine the LR method with artificial drainage; and (iii) salts can be moved away from the root zone to locations in the soil, other than below the root zone, where they are not harmful (called managed accumulation).

Figure 5.3 Reshaping of field drains.

Table 5.2 Estimated Water Application Needed to Leach Salts (Carrow and Duncan, 2011)	
Percent Salt Reduction (%)	Amount of Water Required (cm)
50	15
80	30
90	60

The LR as defined above is simply the ratio of the equivalent depth of the drainage water to the depth of irrigation water (D_{dw}/D_{iw}) and may be expressed as a fraction or as a percentage. Under the foregoing assumed conditions, this ratio is equal to the inverse ratio of the corresponding electrical conductivities, that is:

$$LR = (EC_{iw}/EC_{dw})$$

where EC_{dw} electrical conductivity for drainage water and EC_{iw} electrical conductivity for irrigation water.

For most surface irrigation systems (furrow and flood), irrigation inefficiency (or over-irrigation) generally is adequate to satisfy the LR. However, poor irrigation uniformity often results in salt accumulation in parts of a field or bed. Adding more water to satisfy a LR reduces irrigation efficiency and may result in the loss of nutrients or pesticides and further dissolution of salts from the soil profile (Carrow and Duncan, 2011; Abd El-Hady et al., 2011). Where shallow watertables limit the use of leaching, artificial drainage may be needed. Drainage ditches can be cut in fields below the watertable level to channel away drainage water and allow the salts to leach out. Table 5.2 shows how much water is required to leach salts. Actual salt reduction depends upon water quality, soil texture, and drainage.

5.4 IRRIGATION WATER MANAGEMENT

Salts are most efficiently leached from the soil profile under higher frequency irrigation (shorter irrigation intervals). Keeping soil moisture levels higher between irrigation events effectively dilutes salt concentrations in the root zone, thereby reducing the salinity hazard. Cardon et al. (2014) stated that most surface irrigation systems cannot be controlled to apply less than 75 or 100 mm of water per application and are not generally suited to this method of salinity control. Carrow and Duncan (2011)

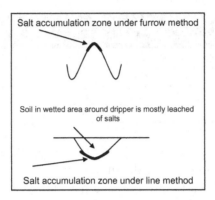

Figure 5.4 Distribution under surface drip irrigation in furrow and line methods (Abou-Baker et al., 2011).

stated that most crop plants are more susceptible to salt injury during germination or in the early seedling stages. An early-season application of good quality water, designed to fill the root zone and leach salts from the upper 15 to 30 cm of soil, may provide good enough conditions for the crop to grow through its most injury-prone stages. The impact of salinity may be minimized by appropriately placing the seeds (or plants) on ridges. Where exactly the seeds should be planted on the ridge or bed will depend on the irrigation design. If the crop planted on ridges would be irrigated via furrows on both sides of the ridge, it is better to put seed plants or seedlings on the ridge shoulders rather than the ridge top because water evaporation will concentrate more salts on the ridge top or center of the bed. Figure 5.4 illustrate that distribution under surface drip irrigation in furrow and line methods.

Sprinkler-irrigated fields with poor water quality present a challenge because it is difficult to apply enough water to leach the salts and you cannot effectively utilize row or bed configurations to manage accumulation. Growers should monitor the soil EC and irrigation water salinity where adequate irrigation water exists above crop requirements (Qadir and Oster, 2004).

5.5 MULCH MANAGEMENT

Klocke et al. (2009) mentioned that crop residue at the soil surface reduces evaporative water losses, thereby limiting the upward movement of salt (from shallow, saline groundwater) into the root zone. Evaporation and thus, salt accumulation, tends to be greater in bare

Figure 5.5 Some materials which used in mulching.

soils. Fields need to have 30−50% residue cover to significantly reduce evaporation. Under crop residue, soils remain wetter, allowing fall or winter precipitation to be more effective in leaching salts, particularly from the surface soil layers where damage to crop seedlings is most likely to occur. Plastic mulches used with drip irrigation effectively reduce salt concentration from evaporation.

Living with salinity by different soil managements like mulch is considering an important practice to sustaining agricultural production. Furrow with mulch practice decreased salinity significantly compared to furrow without mulch. Furrow without mulch method is extensively used in Egypt but it leads to excessive evapotranspiration and increased salinity, especially at the surface layer. This may indicate that mulch practice can contribute to sustaining agriculture through water saving, decreasing evaporation, increasing salt leaching, decreasing salt accumulation, and increasing water intake resulting in a higher yield of sorghum (Al-Dhuhli et al., 2010), wheat (Yan-min et al., 2006), and bean (Abou-Baker et al., 2011). The mulch process exposes the ground to help retain soil moisture and reduce erosion (FAO, 2005). Many materials have been used as mulch, such as plastic film, crop residue, straw, paper pellets, gravel-sand, rock fragment, volcanic ash, poultry and live-stock litters, city rubbish, etc. (Yan-min et al., 2006). Under such land management, at least 30% of the crop residues may remain on the soil surface, for example, straw residues between rows (Ondrasek et al., 2011). Palm leaves mulch was more effective in conserving soil water content, reducing salt accumulation in the soil and resulting in a higher yield of sorghum compared to plastic mulch (Al-Rawahy et al., 2011). Figure 5.5 illustrate some materials which used in mulching.

CHAPTER 6

Crop Management on Salt-Affected Soils

6.1 SELECTION OF TOLERANT CROPS AND ROTATION

The cropping pattern in Egypt is adjusted to soil condition. In the Northern Delta, where soil salinity is somewhat higher than normal, crop rotation includes rice and cotton as the main summer crops and wheat and clover as the main winter crops. All these crops have proven to be tolerant or semi-tolerant to salinity. Sugar beet, which is known to be tolerant to salinity and has been recently grown in the Northern Delta, actively takes up higher amounts of Na^+ and utilizes it for regulation of the osmotic potential in leaves (FAO, 2005). Using crop rotation, minimum tillage, minimum fallow periods, and includes crops in the rotation which ensure effective leaching of accumulated salts (Davis et al., 2012). Sucrose plays a main role in the regulation of the root osmotic potential, followed by potassium, glucose and sodium (Eisa and Ali, 2003). Plant adaptations to salt stress are of three diverse types: osmotic tolerance, sodium chloride exclusion, and the tolerance of tissue to accumulated sodium. The ability of plant cells to keep low cytosolic sodium concentrations is an essential process associated with the capacity of plants to grow in high salt concentrations (Batool et al., 2014).

Excessive soil salinity reduces the yield of many crops. Tolerance of some plants to salinity presents in Table 6.1. This ranges from a slight crop loss to complete crop failure, depending on the type of crop and the severity of the salinity problem. Although several treatments and management practices can reduce salt levels in the soil, there are some situations where it is either impossible or too costly to attain desirably low soil salinity levels (Qadir and Oster, 2004). Some of the most produced and widely used crops in human/animal nutrition such as cereals (rice, maize), forages (clover), or horticultural crops (potatoes, tomatoes) usually require irrigation practices, but are relatively susceptible to excessive concentration of salts either dissolved in

Integrated Management of Salt Affected Soils in Agriculture. DOI: http://dx.doi.org/10.1016/B978-0-12-804165-9.00006-6

Table 6.1 Tolerance of Some Plants to Salinity				
Tolerance Degree	Crop Name	Scientific Name	EC (dS m^{-1})	Salt Rank
Sensitive	Bean	*Phaseolus vulgaris L*	0–4	low
	Carrot	*Daucus carota L.*		
	Onion	*Allium cepa L.*		
	Almond	*Prunus duclis Mill.*		
	Clover	*Trifolium hybridum L.*		
	Potato	*Solanum tuberosum L.*		
	Alfalfa	*Medicago sativa L*		
Moderately sensitive	Muskmelon	*Cucumis melo L.*	4–6	Moderate
	Eggplant	*Solanum melongena L.*		
	Radish	*Raphanus sativus L.*		
	Lettuce	*Lactuca sativa L.*		
	Pepper	*Capsicum annuum L.*		
	Garlic	*Allium sativum L.*		
	Potato	*Solanum tuberosum L.*		
	Cabbage	*Brassica oleracea L.*		
	Celery	*Apium graveolens L.*		
	Spinach	*Spinacia oleracea L.*		
	Cucumber	*Cucumis sativus L.*		
	Tomato	*Lycopersicon esculentum L.*		
	Pea	*Pisum sativum L.*		
	Corn	*Zea mays L.*		
	Green sorghum	*Sorghum bicolor L.*		
	Soybean	*Glycine max L.*		
Moderately tolerant	Red beet	*Beta vulgaris L*	6–8	High
	Fig	*Ficus carica L.*		
	Oats	*Avena sativa L.*		
	Pomegranate	*Punica granatum L.*		
	Sunflower	*Helianthus annuus L.*		
	Wheat	*Triticum aestivum L.*		
Tolerant	Asparagus	*Asparagus officinalis L.*	8–16	Excessive
	Barley	*Hordeum vulgare L.*		
	Cotton	*Gossypium hirsutum L.*		
	Olive	*Olea europaea L.*		
	Rye	*Secale cereale L.*		
	Wheatgrass	*Agropyron cristatum L.*		

(Continued)

Table 6.1 (Continued)				
Tolerance Degree	Crop Name	Scientific Name	EC (dS m^{-1})	Salt Rank
High tolerant	Black greasewood	*Sarcobatus vermiculatus*	Above 16	Very excessive
	Inland saltgrass	*Distichlis stricta*		
	Saltbush species	*Atriplex spp*		
	Alkali bluegrass	*Poa juncifolia*		
	Red glasswort	*Salicornia ruba*		
Source: *Adapted from Maksimovic and Ilin (2012) and FAO (2005).*				

irrigation water or present in soil (rhizosphere) solution. Furthermore, opportunity cropping assumes a wide range of other measures, such as double cropping (cereals and forages), selection of crop species (e.g., perennial deep rooted to maximize water extraction, more salt-tolerant, etc.), presence of pasture and tree species (for windbreaks, controlling groundwater level) (Ondrasek et al., 2011).

6.2 SEEDS PRETREATMENT

Maximum germination of *Chenopodium glaucum* L. was obtained in distilled water. Germination decreased with an increase in salinity. The inhibition of germination by salt solutions was in the order of $MgCl_2 > Na_2SO_4 > Na_2CO_3 > NaCl >$ Soil extract $> MgSO_4$ (Duan et al., 2004).

The pretreatment of seeds with H_2O_2 induced acclimation of the plants to salinity. It decreased the deleterious effects of salt stress on the growth of maize. In addition, the differences in antioxidative enzyme activities may explain the increased tolerance to salt stress of plants originating from H_2O_2 pretreated seeds (Gondim et al., 2010).

Seed germination of halophytes distributed in arid environments is influenced by osmotic and ionic stresses caused due to major variations in water and salinity. Seed pretreatment with H_2O_2 and NaOCl prior to seeding could increase emergence, which may help in mass propagation of *Suaeda fruticosa* (Hameed et al., 2009). The use of various growth regulators did not ameliorate seed germination of *S. fruticosa* except for ascorbic acid (Khan and Gul, 2006). Çavuşoğlu et al. (2013) studied the effects of boric acid pretreatment on seed germination. Boric acid application alleviated in varying degrees the inhibitive effect of salt on the final germination percentage of barley. In 0.30 M salinity, B increased the epidermis cell width on the upper surface, the stomata length on the lower surface, and the epidermis cell number and stomata width on both surfaces. It decreased the epidermis cell width and length on the lower surface, the stomata number and index on both surfaces, and the leaf thickness and distance between vascular bundles. B pretreatment makes water and food transport easy by reducing the distance between vascular bundles in 0.30 M level of NaCl. Moreover, the mentioned application provides adaptation to saline conditions by decreasing the stomata number on both surfaces of the leaf in 0.30 M salinity, and so reduces transpiration and water loss.

Pretreatment with salicylic acid (SA) induced adaptive responses in alfalfa plant under salinity stress and consequently, encouraged protective reactions in biotic membranes which improved the growth of alfalfa seedlings. SA pretreatment improved growth and resulted in higher resistance of plants to salinity, so that it increased germination percentage (Torabian, 2011). The salt tolerance can be improved by many natural, chemical, or biological treatments as described by Dong (2012), Gurmani et al. (2013), and Hussein and Abou-Baker (2014) (Table 6.2).

Table 6.2 Decreasing Salt Stress by Using Some Agents

Agents	Mode of Application	Effects on Stressed Seeds or Plants	Reference
Silicate	Foliar spray	Increasing salt tolerance and promoting bean growth	Abou-Baker et al. (2011)
Salicylic	Foliar spray	Lowering the adverse effect of salt stress and can use for amelioration of salt stress in cotton plants.	Hussein et al. (2012)
Silicate + salicylic	Foliar spray	increasing moringa growth parameters and enhance its mineral status under salt stress	Hussein and Abou-Baker (2014)
Kinetin	Foliar spray	Enhancement effect of growth regulators on chlorophylls, protein content and polyamines titer	Alsokari (2011)
Abscisic acid (ABA)	Foliar spray	the regulation of plant water balance and osmotic stress tolerance	Jonáš et al. (2012)
ABA + Silicate	Foliar spray	Increasing net assimilation rate and stomatal conductance of salt affected rice seedlings	Gurmani et al. (2013)

(Continued)

Table 6.2 (Continued)

Agents	Mode of Application	Effects on Stressed Seeds or Plants	Reference
Cytokinin analog	Seed soaking before sowing plus foliar spray at 45 days after planting	Enhancement of seed germination, seedling growth, and boll setting under salinity	Stark (1991)
	Solution culture of seedling	Leading to recovery of damaged PS II centers	Ganieva et al. (1998)
Coronatine	Applied hydroponically to cotton seedlings at the two leaf stage for 24 h	Improving the antioxidative defense system and radical scavenging activity	Xie et al. (2008)
5-Aminolevulinic acid	Foliar spray	Improvement of salt tolerance through reduction of NaCl uptake	Watanabe et al. (2000)
Glycine betaine	Seed soaking	Alleviation of salt damage	Li et al. (2008)
Potassium (K$^+$)	Foliar spray	Enhancing growth and some chemical constituents of cotton under salt stress	Hussein and Abou-Baker (2014)
Calcium sulfate (Ca^{2+})	Seed soaking before sowing	Enhancement of seed germination, seedling growth under salinity	Javid et al. (2001)
	Applied hydroponically to cotton seedlings	Offsets the reduction in root growth caused by NaCl.	Kent and Läuchli (1985)
VAM (vesicular arbuscular mycorrhizal)	Mixed with soil	Increasing salt tolerance of cotton seedlings	Jalaluddin (1993)
Rs-5 strain (*Klebsiella oxytoca*)	Inoculation through seed soaking	Enhancement of germination and emergence under salinity stress	Yue et al. (2007)
Rs-198 strain (*Pseudomonas putida*)	Inoculation through seed soaking	Protects against salt stress and promote cotton seedling growth	Yao et al. (2010)
Zeolite	Soil application	Ameliorating salinity stress and improving nutrient balance	Al-Busaidi et al. (2007, 2008)

6.3 HALOPHYTES

Another solution used for the recovery of saline soils in agricultural systems or salinized, abandoned areas is the use of plants (Hatton and Nulsen, 1999). The use of plants to remediate saline and sodic soils is a low-cost, emergent method, but with little acceptance due to its low profitability. However, some farmers have improved the salinity condition of their soils by planting salt-tolerant trees or forage shrubs (Silva and Fay, 2012). Various plant species (halophytic plants) grow naturally at the

Desert Holly (*Atriplex hymenelytra*) leaves and bracts

Tamarix aphylla

Figure 6.1 Atriplex *and* Tamarix *trees.*

coast and in salinized areas, and can tolerate saline soils whose salt concentrations can reach those found in ocean water and beyond. The compartmentalizing of the ions in the vacuoles, the accumulation of compatible solutes in the cytoplasm, and the presence of genes for salt tolerance confer salt resistance on these plants (Gorham, 1995). The revegetation of salinized areas with halophytic plants is an example of proactive phytoremediation. The tolerance of halophytes to salt has academic and economic importance (Huchzermeyer and Flowers, 2013). Artemisia (*Artemisia monosperma*) is native wild plant used as an anthelmintic in Egypt (Mohamed et al., 2007). Figure 6.1 shows *Atriplex* and *Tamarix* trees. The littoral vegetation not only protects the shores and provides wood for fuel, fodder, thatching material, and honey for coastal populations but also creates substratum, which provides shelter to a variety of animals. These plants are important for medicinal and ecological values (Rameshkumar and Eswaran, 2013). Screening of available local halophytes for salinity tolerance is of considerable economic value for the

Avicennia marina var. *resinifera*, South Australia

Excreted salt on the underside of a *Avicennia arina* var. *resinifera* leaf

Figure 6.2 Avicennia marina tree *and excreted salt on the underside of* leaf.

utilization of heavy salt-affected lands in coastal tidal-flat areas and other saline areas. Figure 6.2 shows the *Avicennia marina* tree and excreted salt on the underside of a leaf. The order of the relative growth yield in seedling was *Tamarix chinensis* > *Suaeda salsa* > *Salicornia europaea* > *Limonium bicolor* > *Atriplex isatidea* > *Apocynum venetum* > *Phragmites australis* > *Sesbania cannabina* (Xianzhao et al., 2013). Kosová et al. (2013) compared differences observed in salt response of closely related species that differ in salt resistance, and showed that (i) salt-resistant plants display an enhanced constitutive expression of several salt-responsive genes, and (ii) they show few salinity-related disturbances in energy metabolism. Koyro et al. (2013) indicated that increased cellular metabolite concentrations (proline, for instance) are generally not directly related to the abundance of enzymes of the last steps of their catalytic pathways. Gil et al. (2013) reported that soluble carbohydrates not only have roles in the osmotic balance, but are key metabolites and signaling molecules with a "wide variability in the responses to salt stress observed in different species, without any clear quantitative or qualitative general patterns of accumulation of specific sugars or polyalcohols." Ahmed et al. (2013) reported optimal growth of the halophytic grass *Aeluropus lagopoides* to occur at relatively low (26 mM) salt concentration, whilst increasing the salt concentration decreased growth and photosynthesis. A decrease in transpiration helped to minimize Na^+ uptake and this, together with increased secretion from salt glands and the upregulation of membrane transport proteins, enabled *A. lagopoides* to compartmentalize Na^+ at salinities up to 373 mM NaCl and maintain K^+ homeostasis to this external salt concentration. Increased research on the selection of halophytic species which have an economic utilization may enable the rehabilitation and revegetation of salt-affected lands given that the appropriate soil and irrigation management is applied. *Kochia indica* is a highly salt tolerant annual halophytic forage plant grown well in coastal salt marsh in Egypt (Tawfik et al., 2013). Halophytes' tolerance mechanisms according to Parida and Das (2005) are (i) high water use efficiency, (ii) low levels of sodium and chloride ions in cytoplasm, (iii) efficient accumulation of compatible solutes, and (iv) low internal carbon dioxide.

CHAPTER 7

Mineral and Organic Amendments of Salt-Affected Soils

7.1 BALANCED FERTILIZATION

Low productivity of saline soils can be attributed not only to plant toxicity due to the salt or to the damage caused by excessive amounts of soluble salts, but also to low soil fertility. The fertility problems are usually evidenced by a lack of organic matter (OM) and of available mineral nutrients, especially nitrogen and phosphorus. These soils are also usually characterized by a reduction in the activities of some key soil enzymes, such as urease and phosphatase (Yuan et al., 2007), which are associated with biological transformations and the bioavailability of nitrogen and phosphorus. The adverse effect of soil salinity on crops depends both on their tolerance and on other factors which play important roles in the selection of the natural soil microbial flora during salinization, such as soil composition, organic matter, pH, heavy metals, and water and oxygen availability.

Plants develop several mechanisms to induce tolerance to overcome salinity effects. Of the several possible mechanisms to reduce the effects of salinity stress, management of mineral nutrients status of plants can be the most efficient defense system (Nazar et al., 2011).

Using several fertilizers that have high solubility and contribute ions that are not in high demand is recommended, since these ions accumulate in the solution and increase electrical conductivity (EC). Fertilizers are salt based and composed of two parts: ions with a negative charge (anions), and ions with a positive charge (cations). When fertilizers are dissolved in or by water, they break down to ions (Sahin et al., 2011). It is recommended to use fertilizers that are based on almost complete plant nutrients with minimal addition of undesired or less required elements like chlorine, sodium, etc.

Integrated Management of Salt Affected Soils in Agriculture. DOI: http://dx.doi.org/10.1016/B978-0-12-804165-9.00007-8

Salinity decreases the water availability due to high negative pressure that reduces the water uptake and the root pressure driven xylem transport. The solution also contains dissolved nutrients; therefore, their uptake is affected as well. Decreased water uptake reduces the turgor of the leaf cells and thus inhibits the leaf elongation and the cell wall extensibility (El-Shakweer et al., 1998). In a saline medium, both root and shoot growth is depressed, but as a rule shoot growth is more affected. Root elongation is depressed by the presence of high concentrations of NaCl and the low Ca^{2+} concentrations.

The supplementation with nitrogen usually enhances plant growth and yield regardless of whether the plant is salt-stressed or not. Under salt stress condition, nitrogen fertilizer additions mitigate the detrimental effect of salinity on plants (Grattan and Grieve, 1999). Nitrogen use efficiency decreased either by increased salinity or increased nitrogen rates. An apparent increase in salt tolerance was noted when plants were fertilized with an organic nitrogen source compared to that of an inorganic nitrogen source (Huez-López et al., 2011).

A negative linear relationship was observed between gain in plant height, dry matter production, and increasing salinity under both low and increased fertility treatments but the effect was more pronounced under low fertility. All plant characters decreased with increasing salinity levels and were enhanced with increasing fertility (Sharif and Khan, 2009).

Application of inorganic essential nutrients as foliar spray or through the root growing medium has also been reported to be an economical and efficient means of mitigating the adverse effects of salt stress on different crops. Of different major essential nutrients, potassium (K) and phosphorus (P) play vital roles in plant growth and regulate various metabolic reactions (Kaya et al., 2013). Given that excessive Na^+/Cl^- salinity mostly impairs macro/micronutrient balance, the direct practice to recover nutrient uptake and homeostasis is by specific (e.g., Ca, K, P) fertilization. Increasing potassium and phosphorus in the rhizosphere solution could be recommendable for saline conditions, although the upper levels of these nutrient concentrations should be further investigated. Application of zinc (Zn) fertilizer may be beneficial in saline/sodic environment. It was confirmed that $ZnSO_4$ may improve salt tolerance in cereals and results in several other important benefits such as crop micronutrient enrichment and reduced uptake/phytoaccumulation of toxic elements (Ondrasek et al., 2011).

Foliar application with potassium (KNO_3) surpasses all the other treatments (Zn-EDTA and ascorbic acid) especially under high levels of salinity (Tawfik et al., 2013).

Boron is an essential micronutrient for plant growth and development and it is absorbed by plants from the soil solution in the form of boric acid. The combined effects of salinity and excess boron appear to be antagonistic. Broccoli (*Brassica oleracea* L.), however, was more tolerant to boron when the plants were salt-stressed (Smith et al., 2010). Boric acid application alleviated in varying degrees the inhibitive effect of salt on the final germination percentage. Boric acid pretreatment makes water and food transport easy by reducing the distance between vascular bundles in 0.30 M level of NaCl. Moreover, the mentioned application provides adaptation to saline conditions by decreasing the stomata number on both surfaces of the leaf in 0.30 M salinity, and so reduces transpiration and water loss (Çavuşoğlu et al., 2013).

7.2 ORGANIC AMENDMENTS

Recently various organic supplements, such as ground coverings, manures, and compounds, have been investigated for their efficiency in reclaiming saline soils. It has been shown that the application of organic matter can accelerate the leaching of NaCl, decrease ESP and EC, and increase water filtration, the water holding capacity, and aggregate stability (El-Shakweer et al., 1998). The application of decomposing farmyard (FY) manure, straw, or decomposing stable manure significantly increased the productivities of rice, wheat, barley, and sorghum, cultivated in saline soils (Gaffar et al., 1992), the addition of stable manure reduced the sodium adsorption ratio. Tejada et al. (2006) showed that an increase in the organic matter content of saline soils increased the soil structural stability and total porosity and, consequently, the microbial biomass.

The carbon/nitrogen ratio is an extremely important property in the decomposition of organic matter by microorganisms, and for this reason, the organic matter added to saline soils performs an important role in the positive effect on the microbial activity and enzyme activities (Liang et al., 2003). Silva and Fay (2012) added that the use of residues as a soil corrective or conditioner is an economically and environmentally interesting practice, and coconut powder stands out amongst the organic materials that could be used to recover saline soils.

Composted municipal solid waste was commonly used to enhance soil productivity in the agricultural lands and rebuild fertility. However, application of compost on such affected soil helps to diminish salinity thereby improving soil characteristics, mainly by the increase of salts leaching and alleviating the adverse effects of salinity. In addition, high application of compost may be a very useful tool for ameliorating severely salt-affected areas through the establishment of plant cover, including deep-rooted crops (Lakhdar et al., 2009).

Organic fertilizers are considered useful for crops due to their nutritive value, principally of nitrogen, and for their merits in improving the physical properties of the soil, but their salt content is usually ignored, which could prejudice plant growth and soil quality after continued application. The flow of nitrogen and phosphorus from the application of animal manure is considered to contribute to nonprecise pollution (Allen et al., 2006). Eldardiry et al. (2012, 2013) reported that with application of organic manure or humic acid (HA) as a source of organic matter improved soil water movement and hence decreased soil EC and pH which led to an improved yield. Figure 7.1 illustrated that incorporating of organic materials in seed bed.

Agro-materials are associated with stimulation of plant growth, increased crop yield, improved chemical–physical and biological soil conditions, increased nutrient availability, increased soil water retention, and improved quality of crops. Products with these characteristics

Figure 7.1 Incorporating of organic materials in seed bed.

are commercially available to increase production under saline stress conditions (García et al., 2013).

7.2.1 Humic Acid (HA)

The protective effects of HA were shown with common bean plants under salt stress conditions. HA is involved in the metabolic pathways of most plants. It has been shown that HA acts as stimuli for hormonal regulation associated with the metabolism of carbon and nitrogen, antioxidative defensive systems, secondary metabolism, and protective mechanisms against stressors associated with salt and water levels (Aydin et al., 2012). Abiotic stressors from salt content induce the production of reactive oxygen species (ROS) in plants that consequently cause oxidative stress. Application of HA to the roots of rice plants subjected to water stress increased free proline content, protected against membrane permeability, and reduced the H_2O_2 content of the plant tissues. Thus, the potential of these humified materials to exert protective effects against abiotic stressors and improve soil conditions (e.g., number of microorganisms and microbial respiration, soil fertility) have been demonstrated (García et al., 2013).

Application of putrescine and HA positively affected Egyptian cotton plants grown under salt stress, and increased morphological characters, shoot fresh and dry weight. Also, putrescine and HA increased chemical constitutes related to salt tolerance either inorganic, N, P and K, while Na, Cl, Ca and Mg were decreased, or organic constitutes e.g. proline, total free amino acids, total sugars, total soluble phenols, chlorophyll a, b, total chlorophyll and total carotenoids (Ahmed et al., 2013).

7.2.2 Salicylic Acid

Salicylic acid (SA) is a bioregulator of a phenolic nature, and is considered as a hormone. It could be used as a potential growth regulator which can ameliorate the negative effect of salinity and improve plant growth and nutrient utilization under salt stress. Increases in dry weight and number of leaves in salt stressed plants in response to SA may be related to (i) the induction of antioxidant response and the protective role of membranes that increase the tolerance of plant to damage, (ii) SA elongating the cells and increasing cell division and therefore resulting in improving leaf area and stem length (Abbastash et al., 2013), (iii) the exogenous application of SA mitigating the adverse effects of salinity on maize plants by osmoregulation which is

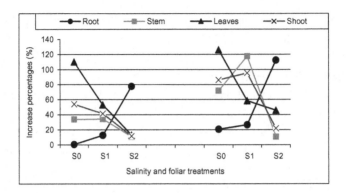

Figure 7.2 The increase percentages (%) of root, stem, leaves, and shoot by addition of Si and (Si + SA) compared to control (Hussein and Abou-Baker, 2014).

possibly mediated by increased production of sugar as well as proline (Fahad and Bano 2012). Figure 7.2 illustrate that the increase percentages (%) of root, stem, leaves, and shoot by addition of Si and (Si + SA) compared to control (Hussein and Abou-Baker, 2014).

SA is produced as part of the hypersensitive response to wounding. It is particularly effective against some virus infections. Some plants produce the volatile methyl salicylate which travels through air to both the affected plant and to neighboring plants as a warning. SA may also be exported to other parts of the plant initiating defense pathogen resistant protein synthesis (Olaiya et al., 2013).

Application of 200 ppm SA showed the highest improvement in growth criteria, uptake of macronutrients, and increased Ca:(K + Na) ratio and can be used for the alleviation of salt stress in cotton plants (Hussein et al., 2012).

7.2.3 Ascorbic Acid

Ascorbic acid (vitamin C: $C_6H_8O_6$) is present in all living plant cells, the largest amounts being usually in the leaves and flowers, that is, in actively growing parts. The supply of ascorbic acid to tomato seedlings might decrease the synthesis of active oxygen species and thereby increase tolerance to salt stress (Shalata and Neumann, 2001). Ascorbic acid acts as an antioxidant, especially when sprayed on plant foliage at the appropriate times. Dewdar and Rady (2013) showed that spraying ascorbic acid at a concentration of 400 mg/L at 40, 60, and 80 days after sowing induced a salinity tolerance in cotton plants cultivated in salt-affected reclaimed soils containing salt concentration up to 7000 ppm, thus leading to favorable growth and consequently a high economic yield. This might be due to (i) an increase of endogenous promoting hormones in the plants, (ii) the stimulation of plant growth and CO_2, as well as microbial activity, and (iii) inducing the synthesis and increasing the intensity of the original protein bands, and causing the appearance of additional new bands in faba bean seeds (Mohsen et al., 2013).

Applying growth regulators, especially ascorbic acid, can modify morphological and physiological characteristics of plants and may also induce better adaptation of the plant to the environment improving the growth and yield. It plays a major role in cell division and cell differentiation (Tawfik et al., 2013).

7.2.4 Amino Acids

Compatible osmolytes (a class of small molecules affecting osmosis) are potent osmoprotectants that play a role in counteracting the effects of osmotic stress. Proline, glycinebetaine, polyamines, and carbohydrates have been described as being effective against salt stress (Hare and Cress, 1997). The addition of $2 \mu M$ thiamine alleviated the effects of NaCl on the protein profile. The results showed that after the addition of thiamine, the 24 kDa protein, which disappeared with NaCl treatment, had been initiated again. Moreover, thiamine treatment stimulated the accumulation of the 20 kDa protein (El-Shintinawy and El-Shourbagy, 2001). The addition of amino acids significantly increased plant height, number of branches, fresh and dry flower yield, chlorophyll a, proline and polyphenol, and carotene, as well as significantly decreased the levels of flavonoid. The interaction between the highest level of salinity and high doses of amino acids gave the best results for plant height and number of branches, as well as fresh and dry weights of flowers. The highest contents of chlorophyll a, b, and carotene were recorded in lower saline soil after spraying with amino acids at 375 ppm (Omer et al., 2013).

The different antioxidants (aspartic acid and glutathione) could partially alleviate the harmful effects of salinity stress that reflect on growth and some physiological changes (increased plant growth, the contents of anthocyanin, α-tocopherol, ascorbic acid and enzymatic activities; in addition, the content of endogenous amino acids was increased which in turn led to positive changes in the picture of protein electrophoresis) of tomato plants (Akladious and Abbas, 2013).

Proline is the most common osmolyte accumulating in plants in response to various stress conditions (Kahlaoui et al., 2013). Proline is a proteinogenic amino acid with an exceptional conformational rigidity, essential for primary metabolism, which normally accumulates in large quantities in response to drought or salinity stress. Its accumulation normally and rapidly occurs in the cytosol where it contributes substantially to the cytoplasmic osmotic adjustment. In addition to its role as an osmolyte for osmotic adjustment, proline contributes to stabilizing subcellular structures (e.g., membranes and proteins), scavenging free radicals, and buffering cellular redox potential under stress conditions (Shanker and Venkateswarlu, 2011). It may also function as a protein compatible hydrotrope, alleviating cytoplasmic acidosis. In addition, rapid breakdown of proline upon relief of stress may provide sufficient reducing agents that support mitochondrial oxidative phosphorylation and generation of ATP for recovery from stress and the repair of stress induced damages (Hare and Cress, 1997).

Exogenous addition of proline was very effective in counteracting the effect of salt. The possible mechanisms for alleviation of salinity stress by proline include: (i) Proline protects the protein turnover machinery against stress-damage and upregulating stress protective proteins (Mohamed et al., 2007). (ii) Proline led to an increase of leaf area, growth length, and fruit yield. With regards to mineral nutrition, Ca^{2+} was higher in different organs while low accumulation of Na^+ occurred. However, Cl^- was very low significantly in all tissues of tomato plants at the higher concentration of proline applied (Kahlaoui et al., 2013). (iii) Proline may serve as a nitrogen and carbon source needed in stress recovery and may also act as a scavenger of hydroxyl radicals, helping the plant to avoid cellular damage provoked by osmotic or salt-induced oxidative stress (Kahlaoui et al., 2013).

7.2.5 Indoleacetic Acid (IAA)

Exogenous application of different types of chemicals including plant growth regulators, osmoprotectants, and inorganic nutrients seems to be an efficient, economical, and shot-gun approach to adverse effects of salt stress on plants. The use of such substances has resulted in a substantial increase in both growth and yield of many crops grown under saline conditions (Ashraf et al., 2008).

Of the various plant growth regulators which regulate growth under normal or stress conditions, indoleacetic acid (IAA), one of the key auxins occurring naturally in plants, plays a vital role in maintaining plant growth under stress conditions including salt stress. Foliar application of IAA (15 mg/L) considerably ameliorated the adverse effects of salt on these plants (Guru Devi et al., 2012).

Although exogenous application of potassium, phosphorus or IAA proved to be effective in alleviating the adverse effects of saline stress on growth and kernel yield of maize plants, application of these chemicals when applied in combination had a pronounced effect in terms of promoting growth and kernel yield of salt stressed maize plants, reducing membrane permeability, enhancing photosynthetic pigment concentration, and reducing the leaf Na^+/K^+ ratio under saline medium (Kaya et al., 2013).

7.2.6 Phytohormones

Phytohormones affecting the impact of stress on plants were used for this purpose.

Abscisic acid (ABA) is an important phytohormone and plays a critical role in response to various stress signals. As many abiotic stresses ultimately result in desiccation of the cell and osmotic imbalance, the main function of ABA seems to be the regulation of plant water balance and osmotic stress tolerance. Abscisic acid, 24-epibrassinolide, kinetin, and spermine were applied by spraying the leaf in three concentrations. The results showed that there were highly significant differences compared to controls with other variants, especially in the evaluation of physiological parameters. The most significant influence on the stomatal conductance was observed in the variants treated with abscisic acid (Jonáš et al., 2012). Exogenous application of kinetin and spermine mitigated the deleterious effects of salinity stress on growth and yield of the plants. Conversely, the combined treatment of kinetin and spermine induced additional reduction in growth and yield of the stressed plants, and the effect appeared to be constitutive. The protective effect of kinetin and spermine on *Vigna sinensis* plants appeared mainly due to the enhancement effects of these growth regulators on chlorophylls, protein content and polyamines titer (Alsokari, 2011). Under saline conditions, ABA, Si and ABA + Si application ameliorated plant growth via suppression of Na^+ accumulation in shoots and lowering the

Na^+/K^+ ratios. In addition, treatment with Si alone or with ABA significantly reduced Na^+ concentrations in the leaf blades and sheaths, increased the net assimilation rate, and the stomatal conductance of salt-affected rice seedlings (Gurmani et al., 2013).

7.2.7 Glucosinolates

Glucosinolates (a class of secondary metabolites) are nitrogen- and sulfur-containing compounds mainly found in *Capparales* and almost exclusively in *Brassicaceae*, which include Brassica crops of economic and nutritional importance, as well as the model plant *Arabidopsis thaliana*. Martínez-Ballesta et al. (2013) reported that the exogenous addition of glucosinolate hydrolysis products may alleviate certain stress conditions through its effect on specific proteins. The function of glucosinolates, further than in defense switching, is alleviating abiotic stress (salinity stress). The fact that a transient allocation and redistribution of glucosinolates in response to environmental changes is observed could give an indication that glucosinolate-specific function under abiotic stress is still unclear and requires further attention.

7.3 SILICATES

Although silicon (Si) is the second most abundant element both on the surface of the Earth's crust and in soils, it has not yet been listed among the essential elements for higher plants. However, the beneficial role of silicon in stimulating the growth and development of many plant species has been generally recognized. Silicon is known to effectively mitigate various abiotic stresses (Liang et al., 2007). The physically and chemically active silicon substances in the soil are represented by soluble monosilicic acids, polysilicic acids, and organosilicon compounds. The soluble monosilicic acids are absorbed by plants and microorganisms. Polysilicic acid has a significant effect on soil texture, water holding capacity, adsorption capacity, and stability of soil erosion. Plants can absorb enough silicon, which can determine silicon effect on the soil fertility and plants. Silicon is absorbed by plants in the form of uncharged silicic acid, $Si(OH)_4$, and is ultimately irreversibly precipitated throughout the plant as amorphous silica. Increasing silicon content in plant tissues enhances their resistance to various stresses. The presence of silicon in the cell walls of plants increases their strength, as silicon increases resistance to salinity

(Balakhnina and Borkowska, 2013). Silicon interacts with polyphenols in xylem cell walls and affects lignin deposition and biosynthesis.

Several functions have been attributed to silicon: improvement of nutrient imbalance, reduction of mineral toxicities (heavy metal mobility), improvement of mechanical properties of plant tissues, controlling chemical and biological properties of the soil, microbial activity, stability of soil organic matter and formation of polysilicic acids in the soil, and enhancement of resistance to other various abiotic (salt, metal toxicity, nutrient imbalance, drought, radiation, high temperature, freezing, UV) and biotic stresses (Ma and Yamaji, 2006; Liang et al., 2007). Salinity significantly decreased the fresh and dry weights of shoot and root, stem length, leaf area, chlorophyll content, and relative water content of maize plants and an application of silicon significantly increased them. Silicon increased the physiological properties of maize; therefore, proper silicon nutrition can increase salt resistance in maize plants (Rohanipoor et al., 2013). Silicon application significantly increased 100 seed weight and yield of bean under saline environments (Parande et al., 2013). Abou-Baker et al. (2012) reported that the highest basic branch, bods number, biological yield, stover yield, and plant height were obtained from potassium silicate application under salt stress conditions (Figure 7.3) and confirmed the pathway which confirmed silicon as one of the essential elements to plant growth.

The different mechanisms by which silicon stimulates growth and alleviates salt stress in plants are described in Zhu et al. (2011), Ashraf et al. (2011), Abou-Baker et al. (2011), and Balakhnina and

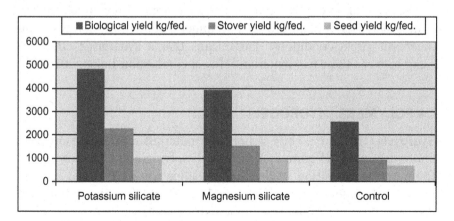

Figure 7.3 Effect of potassium and magnesium silicate on bean yield.

Borkowska (2013): (i) Decreased mutual shading through improving leaf erectness. (ii) External silicon enrichment not only reduced Na uptake and accumulation but also influenced its distribution in plant parts and consequently improved the capability for adaptation to salinity stress. (iii) Silicon application improved potassium uptake and enhanced K:Na selectivity ratio, which mitigated against the toxic effects of sodium. (iv) Addition of silicon decreased the permeability of plasma membrane of leaf cells and significantly improved the photosynthetic activity and ultra-structure of chloroplasts which were badly damaged by the added NaCl. (v) Silicon application increased leaf chlorophyll content and plant metabolism and mitigated nutrient imbalance and metal toxicity in plants. (vi) Silicon application led to an improvement in plant water status and storage within plant tissues. (vii) Silicon application could mitigate the inhibitory effect of salinity on net photosynthesis and this effect was associated with lower Cl translocation. (viii) Silicon was deposited in the cell walls of roots, leaves, stems, and hulls and modified the cell wall architecture, reducing the translocation of salts to the shoots. (ix) Silica deposition in the leaf limited transpiration, and the partial blockage of the transpirational bypass flow. (x) Silicon reduced ROS generation, intensity of lipid peroxidation, and in some cases, increased the activity of enzymes of ROS detoxificators: superoxide dismutase, ascorbate peroxidase, glutathione reductase, guaiacol peroxidase, and catalase. (xi) Stimulation of antioxidant systems in plants. This review describes the progress in the roles of silicon-mediated alleviation of salt stress in higher plants and its important role in enhancing plant salt stress tolerance.

The application of potassium silicate in combination with SA gave the highest increases in plant growth and mineral content of moringa plants under salt stress conditions. This means that a synergistic effect was found between these two materials (Hussein and Abou-Baker, 2014).

7.4 POLYMERS (HYDROGEL)

Hydrogels (HG) are a class of polymer materials that can absorb large amounts of water (Figure 7.4) without dissolving and they vary in their origin and composition. HG added to saline soil in arid and semiarid areas significantly improved the variables affected by high salinity and also reduced soil electricity conductivity, electrolyte leakage of plant, and growth parameters. HG appear to be highly effective for use as soil conditioners in vegetable growing, and to improve crop tolerance and

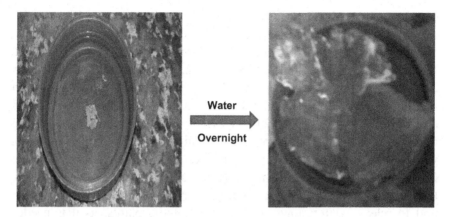

Figure 7.4 Hydrogel can absorb large amounts of water.

growth in saline conditions (Kant and Turan, 2011). Thus, the use of hydrophilic polymer in soils, especially in sandy soils, increases soil water holding capacity, yield, and water use efficiency of plant (El-Hady and Abou-Sedera, 2006b). Furthermore, HG decrease the negative effect of soil salinity on plants and help irrigation projects to succeed in arid and semiarid areas (Dorraji et al., 2010). HG caused the residual water content and saturated water content to increase. Available water content increases to a maximum of about 2.3 times the control. Application of HG can result in significant reduction in the required irrigation frequency, particularly for coarse-textured soils. The enhancement of the water management of coarse-textured soils is an important issue in arid and semiarid regions of the world. HG effects on growth and ion relationships of salt resistant woody species, *Populus euphratica*. The addition of 0.6% HG to saline soil improved seedling growth (2.7-fold higher biomass) during a period of 2 years. Root length of plant grown in treated soil with HG had 3.5 fold more than those grown in untreated soil. Hydrogel treatment enhanced Ca^{2+} uptake and increased capacity of *Populus euphratica* to exclude salt (i.e., reduces contact with Na^+ and Cl^-) (Koupai et al., 2008).

7.5 ZEOLITE

Zeolites are microporous, aluminosilicate minerals, zeolites occur naturally but are also produced industrially on a large scale. In general, three important factors, structure, texture, and chemical composition, as well as the economic value of natural and synthetic zeolites have made them valuable materials (Ghorbani and Babaei, 2008).

Total world production of natural zeolites is between 3 Mt and 4 Mt per year. Estimates for individual countries are China, 2.5 Mt; Cuba, 500,000 to 600,000 t; Japan, 140,000 to 160,000 t; USA, 42,000 t; Hungary, 10,000 to 20,000 t; Slovakia, 12,000 t; Georgia, 6,000 t; New Zealand, 5,000 t; Greece, 4,750 t; Australia, Canada, Italy, and other republics of the former Soviet Union, 4,000 t each; Bulgaria, 2,000 t; and South Africa 1,000 to 2,000 t (Virta, 2000). According to reports of 2001, the total consumption of zeolites was 3.5 million tons of which 18% came from natural resources and the rest from synthetics (Polat et al., 2004). Due to their unique structure, zeolites are able to trap liquids from excessive irrigation or rain or from liquid fertilizers and to release water slowly to the roots of the plants, as it is needed (Mohammad et al., 2004). Increasing the rate of mixed zeolite into sandy soil increased the soil moisture content, reduced the rate of nitrification likely because of NH_4^+ adsorption in the zeolite mineral lattice, and may reduce the leaching of inorganic nitrogen (Ippolito et al., 2011). This may increase salt tolerance by plants. Zeolites assist water infiltration and retention in the soil due to their very porous properties and the capillary suction they exerts. In this concern Al-Busaidi et al. (2007, 2008) reported that zeolite application significantly increased water holding capacity of the soil and accumulated more salts. The zeolite mixed soils improved plant growth compared to the un-amended control. Soil analysis showed high concentrations of Ca^{2+}, Mg^{2+}, Na^+, and K^+ due to saline water especially in the upper soil layer but concentrations were lower in soils treated with zeolites. Soil amended with zeolites could effectively ameliorate salinity stress and improve nutrient balance in a sandy soil. Yasuda et al. (1998) reported that the zeolite has an effect of mitigating the salt damage to plants and these mitigation effects are the result of a reaction exchanging sodium in salt water for calcium from the zeolite. Although, zeolite could probably reduce the electrical conductivity of soil solutions by adsorption of ions from the primary solutions, it seems that zeolites would tend to adsorb more potassium ions compared to sodium ions from the solutions resulting in a lower electrical conductivity of potassium containing zeolites compared to sodium (Ghorbani and Babaei, 2008). The addition of the K-type and Ca-type artificial zeolites ameliorated beet growth by the increase of the supply and uptake of potassium and calcium, the decrease of nitrogen concentration, and the improvement of the cation balance. The K-type zeolite exhibited an adequate K-supplying ability (Yamada et al., 2002). In addition, Ashraf (2011) suggested that the addition of zeolite in perlite and pumice could improve inorganic medium properties for tomato soilless culture, leading to higher yields. Medium amendment with zeolite could effectively ame-liorate salinity stress and improve nutrient balance in the medium. In another study, the greatest improvement in sandy soil (EC = 17 dS/m) characteristics was recorded under the highest level of zeolite and benton-ite mixture (2.5 ton/fed.). Moreover, growth parameters of the two crops (bean and corn) as well as their seed yields were beneficially increased by increasing the rate of applied conditioner. This reveals the great advan-tage of combining natural zeolite and bentonite conditioners in a proper amount to improve sandy soils properties and enhance their productivity (Hassan and Mahmoud, 2013).

7.6 HALOPHILIC MICROORGANISMS

7.6.1 Bacteria

Halophilic bacteria are adapted to high osmolarity, can grow in high saline environments, and are commonly found in natural environments containing significant concentration of NaCl (Irshad et al., 2013).

Sulfur inoculated with *Thiobacillus* was more efficient than gypsum in the reduction of the exchangeable sodium of the soil and promoting leaching of salts, especially sodium. Sulfur inoculated with *Thiobacillus* reduced the EC of the soil saturation extract to levels below that adopted in soil classification of sodic or saline sodic (Stamford et al., 2002).

Soil inoculations with the bacterial cells of *Azotobacter* (AZ), Streptomycin (ST) or AZ + ST increased the growth and development of wheat in normal and saline conditions. This may be due to increas-ing root depth, shoot length, shoot and root dry weights, levels of phosphorus, nitrogen, magnesium, potassium, and proteins present in wheat shoots, and decreasing proline concentration (Aly et al., 2012). Dhanraj (2013) isolated 43 bacterial strains from saline soil, and showed that AZ and *Bacillus subtilis* were found to be the most domi-nant nitrogen-fixing bacteria and phosphate solubilizer, respectively. In conclusion, results of these studies indicated that the importance of these organisms as bioinoculants for saline soil and can be explored for biofertilizer production.

Plant growth promoting rhizobacteria (PGPR) have the ability to reduce the deleterious effect of salinity on plants due to the presence of ACC-deaminase enzyme (Ahmad et al., 2014) along with mechan-isms such as nitrogen fixation, solubilization of phosphorus, increasing growth by changing the endogenous level of plant growth regulating substances, or indirectly by increasing the natural resistance of the host against pathogens. Some PGPR strains have the ability to lower the level of ethylene in stressed plants and eliminate/reduce the potential inhibitory effect of high ethylene concentrations through development of better root systems, thus helping to alleviate the stressed conditions in plants (Shaharoona et al., 2006; Ahmad et al., 2014).

7.6.2 Fungi

Arbuscular Mycorrhiza fungi (AMF) have been shown to promote plant growth, nutrient uptake (particularly important for phosphorus) and reduce salt stress in plant. The fresh and dry biomass of onion plants is higher in AMF inoculated plants than noninoculated (Shinde et al., 2013). The mycorrhizal inoculation is capable of alleviating the damage caused by salt stress conditions on pepper plants, by maintaining the membranes stability and plant growth. Mycorrhizal symbiosis is a key component in helping plants to cope with adverse environmental conditions by (i) improvement of plant nutrient uptake, particularly P, (ii) elevation of K:Na ratio, (iii) providing higher accumulation of osmosolutes, (iv) maintaining higher antioxidant enzymatic activities, (v) increasing efficiency of photosystem, stomatal conductance, and glutathione content and by reducing oxidative damage (Estrada et al., 2013), (vi) higher ability for taking up the immobile nutrients in nutrient-poor soils as well as improvement of tolerance to salinity, and (vii) some aquaporin genes are upregulated in mycorrhizal plants, causing significant increase in the water absorption capacity of salt-affected plants (Hajiboland, 2013).

7.6.3 Yeast

Gomaa and Gaballah (2004) studied the effect of salt stress on four Egyptian maize varieties grown under different salinity levels in relation to biofertilization with yeast (*Rhodotorula glutinis*). The results showed that biofertilization alleviated adverse effects of high levels of salinity and plants accumulated more polyamines than those that received no biofertilizer, especially at high salinity levels. This increment may be due to (i) increasing total soluble proteins content, which could be attributed to the growth hormones produced by yeast (Gaballah and Gomaa, 2004), (ii) application of the yeasts significantly increased the photosynthetic pigments, soluble sugars, sucrose, and total soluble proteins of sugar beet—it increased the sucrose content by about 43% of the control, (iii) the anatomy of the leaf and the root showed an increase in thickness of the blade, mid vein, dimensions of the vascular bundles, and number and diameter of xylem vessels as the result of the application of yeasts (Agamy et al., 2013).

7.6.4 Azolla

Azolla is a free floating water fern that floats in the water and fixes atmospheric nitrogen (Figure 7.5) because of its association with the nitrogen-fixing cyanobacterium *Anabaena*. Several studies have indicated that *Azolla* and cyanobacteria as biofertilizer could improve plant growth and yield under salinity conditions (Waseem et al., 2012). *Azolla pinnata* var. pinnata and Rong Ping were the best species to be used as an inoculum for the saline soils in a rice wheat cropping system. The two- to four-fold growth of some *Azolla* species due to P application indicated that significantly higher *Azolla* biomass can be produced in saline soils, by just application of P fertilizer in saline soils (Hamid et al., 2007).

Application of *Azolla* as a biofertilizer with biostraw significantly increased nitrogen, phosphorus, and potassium content of rice grains and straw; moreover a combination of *Azolla* with cyanobacteria and biostraw increased the soil organic matter and nitrogen, phosphorus, and potassium (available in soil and plant content), slightly decreased soil EC, showed several benefits over chemical fertilizers, and improved the fertility of saline soils (Abd El-All et al., 2013).

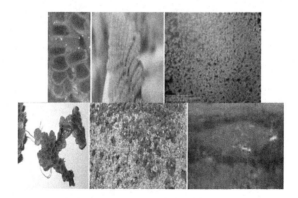

Figure 7.5 Azolla *growth.*

CHAPTER 8

Genetic Engineering and Economic Aspects

8.1 GENETIC ENGINEERING

The mechanisms of genetic control of salt tolerance in plants have not yet fully understood because of its complexity. There are in fact several genes controlling salinity tolerance in the different species whose effect interacts strongly with environmental conditions. Thus, genetic variation can only be demonstrated indirectly, by measuring the responses of different genotypes. Plants have evolved several mechanisms to acclimatize to salinity. It is possible to distinguish three types of plant response or tolerance: (i) the tolerance to osmotic stress; (ii) the Na^+ exclusion from leaf blades; and (iii) tissue tolerance (Shanker and Venkateswarlu, 2011). There are many proteins that appear related to salt stress response in plants. These proteins can directly regulate the levels of osmolytes (mannitol, fructans, proline, and glycine betaine) and control ion homeostasis. Genetic engineering of these osmolytes resulted in increasing salt tolerance. Molecular and transgenic breeding is more expensive than conventional breeding, but represents an efficient way to produce salt-tolerant lines (Ondrasek et al., 2011).

Use of both genetic manipulation and traditional breeding approaches will be required to unravel the mechanisms involved in salinity tolerance and to develop salt-tolerant cultivars better able to cope with the increasing soil salinity constraints. Through a combination of such master genes that act in different pathways reactive oxygen species scavenging and osmotic adaptation may prove even more beneficial for improving stress tolerance (Batool et al., 2014). Most of the research and field trials in Egypt are being carried out to enhance osmotic stress tolerance in wheat and tomato by transferred osmoregulatory genes from barley and yeast (Mohamed et al., 2007).

Integrated Management of Salt Affected Soils in Agriculture. DOI: http://dx.doi.org/10.1016/B978-0-12-804165-9.00008-X

8.2 ECONOMIC ASPECTS

When the accumulated salt in soil layers is above a level that adversely affects crop production, choosing salt-tolerant crops and managing soil salinity are considered important strategies to boost agricultural economy (Rengasamy, 2010). Processes such as seed germination, seedling growth and vigor, vegetative growth, flowering, and fruit set are adversely affected by high salt concentration, ultimately causing diminished economic yield and also quality of production (Sairam and Tyagi, 2004). Salinity can be minimized with irrigation and drainage, but the cost of engineering and management is very high (Tuna et al., 2008). The accumulation of salts from improper soil and water management is a serious problem worldwide. The rehabilitation cost in Cuba has been estimated at US$ 6000 per hectare for severely affected land, US$ 2000 per hectare for strongly saline land, US$ 1000 per hectare for moderately saline land, and US$ 500 per hectare for weakly saline land (González Núñez et al., 2004). The global cost of irrigation-induced salinity is equivalent to an estimated US$ 11 billion per year (FAO, 2005). Accordingly, in response to the salinity issue, Australia's National Action Plan for Salinity and Water Quality from 2000, resulted in investments of about AU$1.4 billion over 7 years to support actions by communities and land managers in salt-affected regions (Williams, 2010).

Typically, global positioning systems (GPS) range in price from a few hundred to many thousands of dollars or more, depending on the system accuracy specifications. Most types of salinity survey work require survey location accuracies of ±1 to 2 m, which can be obtained from GPS units which facilitate differential correction (the typical cost of a differentially correctable GPS unit starts at around US$ 1000) as described by Rhoades et al. (1999).

Exogenous application of inorganic essential nutrients as foliar spray or through the root growing medium has also been reported to be an economical and efficient means of mitigating the adverse effects of salt stress on different crops (Ashraf et al., 2008). The use of chemical materials such as sodium silicate or potassium silicate as a source of silicon for combating salinity is not economical, while crop residues such as stalks of rice, sugarcane and bagasse (sugarcane pulp) can be used as a source of silicon (Rohanipoor et al., 2013).

SUMMARY AND CONCLUSIONS

Arid and semiarid regions of the world are generally associated with high population density and lower than average per capita incomes and living standards. These regions are vulnerable to food shortages due to the current, unsustainable use of land affected by soil salinization. It is worth mentioning that most of the new reclaimed areas in Egypt are salinity affected.

Salt-affected soils are the main challenge that farmers and researchers face. Such soils are poor with respect to their physico-biochemical properties, and soil water–plant relationships, as well as their nutritional status. Reclamation and land utilization of such soils face several difficulties. To overcome the negative effects of salinity on the plant growth and yield can be to attempt new strategies.

Management of salt-affected soils or living with salinity requires combinations of agronomic practices. It can be reclaimed by (i) scraping and removal of top soil, (ii) land leveling, (iii) subsoiling and deep ploughing, (iv) usage of furrow or bed with mulch cultivation method, (v) adapted irrigation and drainage system, (vi) salt leaching with the addition of a calcium source, (vii) cropping selection and rotation, (viii) use of halophytes, (ix) incorporating organic matter, (x) balanced fertilization, (xi) usage of chemical and organic solutions combination (e.g., humic, ascorbic, amino acids, phytohormones, salicylic and silicate, etc.), (xii) use of polymers (hydrogel), (xiii) finding the most suitable and economical mixture of zeolite in combination with organic fertilizers, compost, and biogases, in addition to, (xiv) incorporating of halophilic microorganisms on applied soil amendments. Periodic soil testing combined with proper management procedures, will improve the adaptation capability of plants to a saline environment.

Generally salt-affected soil management by increasing the plant tolerance (by amendments) and resisting salinity (by breeding and genetic engineering) is more effective in practical and economic aspects.

This brief review of publications about the management of salt-affected soils shows different techniques for living with salinity. However, the relative efficiency of these collected recommendations is not well documented. Therefore, it will need other studies to improve the integrated recommendation system.

REFERENCES

Abbastash, R., Maftoon, M., Zadehbagheri, M., Rousta, M.J., 2013. The effects of seed priming with salicylic acid on the growth of maize under salinity conditions. Int. J. Agric. Crop Sci. 5 (16), 1820–1826.

Abd El-All, A.A.M., Elsherif, M.H., Shehata, H.S.H., El-Shahat, R.M., 2013. Efficiency use of nitrogen, biofertilizers and composted biostraw on rice production under saline soil. J. Appl. Sci. Res. 9 (3), 1604–1611.

Abd El-Hady, M., Shaaban, S.M., 2010. Acidification of saline irrigation water as a water conservation technique and its effect on some soil properties. Am. Eurasian J. Agric. Environ. Sci. 7 (4), 463–470.

Abd El-Hady, M., Eldardiry, E.I., Selim, E.M., 2011. Effect of water regime on some hydrophysical properties and yield component of vine grape under shallow water table soils. J. Appl. Sci. Res. 8 (1), 84–93.

Abdelhamid, M., Eldardiry, E., Abd El-Hady, M., 2013. Ameliorate salinity effect through sulphur application and its effect on some soil and plant characters under different water quantities. Agric. Sci 4 (1), 39–47.

Abou-Baker, N.H., Abd-Eladl, M., Abbas, M.M., 2011. Use of silicate and different cultivation practices in alleviating salt stress effect on bean plants. Aus. J. Basic Appl. Sci. 5 (9), 769–781.

Abou-Baker, N.H., Abd-Eladl, M., Eid, T.A., 2012. Silicon and water regime responses in bean production under soil saline condition. J. Appl. Sci. Res. 8 (12), 5698–5707.

Agamy, R., Hashem, M., Alamri, S., 2013. Effect of soil amendment with yeasts as bio-fertilizers on the growth and productivity of sugar beet. Afr. J. Agric. Res. 8 (1), 46–56.

Ahmad, M., Zahir, Z.A., Jamil, M., Nazli, F., Latif, M., Akhtar, M.F., 2014. Integrated use of plant growth promoting rhizobacteria, biogas slurry and chemical nitrogen for sustainable production of maize under salt-affected conditions. Pak. J. Bot. 46 (1), 375–382.

Ahmed, M.Z., Shimazaki, T., Gulzar, S., Kikuchi, A., Gul, B., Khan, M.A., et al., 2013. The influence of genes regulating transmembrane transport of Na^+ on the salt resistance of *Aeluropus lagopoides*. Funct. Plant Biol. 40, 860–871.

Akladious, S.A., Abbas, S.M., 2013. Alleviation of seawater stress on tomato by foliar application of aspartic acid and glutathione. J. Stress Physiol. Biochem. 9 (3), 282–298.

Alberta Environment, 2000. Interim Salt Contamination Assessment and Remediation Guidelines. Environmental Sciences Division, Environmental Service, Edmonton, Alberta, p. 95.

Al-Busaidi, A., Yamamoto, T., Irshad, M., 2007. The ameliorative effect of artificial zeolite on barley under saline conditions. J. Appl. Sci. 7, 2272–2276.

Al-Busaidi, A., Yamamoto, T., Inoue, M., Eneji, A.E., Mori, Y., Irshad, M., 2008. Effects of zeolite on soil nutrients and growth of barley following irrigation with saline water. J. Plant Nutr. 31 (7), 1159–1173.

Al-Dhuhli, H.S., Al-Rawahy, S.A., Prathapar, S., 2010. Effectiveness of mulches to control soil salinity in sorghum fields irrigated with saline water. A monograph on management of salt-affected soils and water for sustainable agriculture. Sultan Qaboos University, pp. 41–46.

Allen, S.C., Fair, V.D., Graetz, D.A., Shibu, J., Ramachandran, N.P.K., 2006. Phosphorus loss from organic versus inorganic fertilizers used in alley cropping on a Florida Ultisol. Agric. Ecosyst. Environ. 117 (4), 290–298.

Al-Rawahy, S.A., Al-Dhuhli, H.S., Prathapar, S.A., Abdel Rahman, H., 2011. Mulching material impact on yield, soil moisture and salinity in saline-irrigated sorghum plots. Int. J. Agric. Res. 6, 75–81.

Alsokari, S.S., 2011. Synergistic effect of kinetin and spermine on some physiological aspects of seawater stressed *Vigna sinensis* plants. Saudi J. Biol. Sci. 18 (1), 37–44.

Aly, M.M., El-Sayed, H.E.A., Jastaniah, S.D., 2012. Synergistic effect between *Azotobacter vinelandii* and *Streptomyces* sp. isolated from saline soil on seed germination and growth of wheat plant. J. Am. Sci. 8 (5), 667–669.

Ashraf, S., 2011. The effect of different substrates on the vegetative, productivity characters and relative absorption of some nutrient elements by the tomato plant. Adv. Environ. Biol. 5 (10), 3091–3096.

Ashraf, M., Athar, H.R., Harris, P.J.C., Kwon, T.R., 2008. Some prospective strategies for improving crop salt tolerance. Adv. Agron. 97, 45–110.

Ashraf, M., Saqib, R.M., Afzal, M., Tahir, M.A., Shahzad, S.M., Imtiaz, M., 2011. Silicon nutrition for mitigation of salt toxicity in sunflower (*Helianthus annuus* L). In: International Conference on Silicon in Agriculture, Proceedings of the Fifth, September, Beijing, China, p. 3.

Aydin, A., Kant, C., Turan, M., 2012. Humic acid application alleviates salinity stress of bean (*Phaseolus vulgaris* L.) plants decreasing membrane leakage. Afr. J. Agric. Res. 7, 1073–1086.

Azam, F., Ifzal, M., 2006. Microbial populations immobilizing NH_4^+-N and NO_3^--N differ in their sensitivity to sodium chloride salinity in soil. Soil Biol. Biochem. 38 (8), 2491–2494.

Azevedo Neto, A.D., Gomes-Filho, E., Prisco, J.T., 2008. Salinity and oxidative stress. In: Khan, N.A., Sarvajeet, S. (Eds.), Abiotic Stress and Plant Responses. IK International, New Delhi, pp. 58–82.

Balakhnina, T., Borkowska, A., 2013. Effects of silicon on plant resistance to environmental stresses: review. Int. Agrophys. 27, 225–232.

Barroso, C., Romero, L.C., Cejudo, F.J., Vega, J.M., Gotor, C., 1999. Salt-specific regulation of the cytosolic *O*-acetylserine(thiol)lyase gene from *Arabidopsis thaliana* is dependent on abscisic acid. Plant Mol. Biol. 40, 729–736.

Batool, N., Shahzad, A., Ilyas, N., Noor, T., 2014. Plants and salt stress. Int. J. Agric. Crop Sci. 7 (9), 582–589.

Blanco, H., Lal, R., 2010. Principles of Soil Conservation and Management. Springer, Dordrech, Heidelberg, London, New York.

Bray, E.A., Bailey-Serres, J., Weretilnyk, E., 2000. Responses to abiotic stresses. In: Buchanan, B.B., Gruissem, W., Jones, R.L. (Eds.), Biochemistry and Molecular Biology of Plants. ASPP, Rockville, MD, pp. 1158–1203.

Cardon, G.E., Davis, J.G., Bauder, T.A., Waskom, R.M., 2014. Managing Saline Soils. Fact sheet no. 0.503, Colorado State Univ. Extension. <http://www.ext.colostate.edu/pubs/crops/00503.html>.

Carrow, R.R., Duncan, R.N., 2011. Best management practices for saline and sodic turfgrass soils assessment and reclamation. In: Irrigation System Design and Maintenance for Poor-Quality Water, pp. 195–218 (Chapter 10).

Çavuşoğlu, K., Kaya, F., Kılıç, S., 2013. Effects of boric acid pretreatment on the seed germination, seedling growth and leaf anatomy of barley under saline conditions. J. Food Agric. Environ. 11 (2), 376–380.

Charlton, W.L., Matsui, K., Johnson, B., Graham, I.A., Ohme-Takagi, M., Baker, A., 2005. Salt-induced expression of peroxisome-associated genes requires components of the ethylene, jasmonate and abscisic acid signalling pathways. Plant Cell Environ. 28, 513–524.

Davis, J.G., Waskom, R.M., Bauder, T.A., 2012. Managing Sodic Soils. Fact sheet, Colorado State Univ. Extension. <www.ext.colostate.edu/pubs/crops/00504.html/>.

De Kok, L.J., Oosterhuis, F.A., 1983. Effect of frost-hardening and salinity on glutathione and sulfhydryl levels and on glutathione reductase activity in spinach leaves. Physiol. Plant 58, 47−51, In: Nazar et al., 2011, Environ. Exp. Bot, 70, 80−87.

Dewdar, M.D.H., Rady, M.M., 2013. Induction of cotton plants to overcome the adverse effects of reclaimed saline soil by calcium paste and ascorbic acid applications. Acad. J. Agric. Res. 1 (2), 017−027.

Dhanraj, N.B., 2013. Bacterial diversity in sugarcane (Saccharum officinarum) rhizosphere of saline soil. Int. Res. J. Biol. Sci. 2 (2), 60−64.

Dobrovol'skii, G.V., Stasyuk, N.V., 2008. Fundamental work on saline soils of Russia. Eurasian Soil Sci.100−101, Pleiades Publishing, Ltd. In: Lakhdar et al., 2009.

Dong, H., 2012. Technology and field management for controlling soil salinity effects on cotton. Aust. J. Crop Sci. 6 (2), 333−341.

Dorraji, S.S., Golchin, A., Ahmadi, S., 2010. The effects of hydrophilic polymer and soil salinity on corn growth in sandy and loamy soils. Clean—Soil, Air, Water 7 (38), 584−591.

Duan, D., Liu, X., Khan, M.A., Gul, B., 2004. Effects of salt and water stress on the germination of Chenopodium glaucum L. Seed. Pak. J. Bot. 36 (4), 793−800.

Eisa, S.S., Ali, S.H., 2003. Biochemical, physiological and morphological responses of sugar beet to salinization [online]. Zu finden in <http://geb.uni-giessen.de/geb/volltexte/2003/1235/>.

El-Baky, A., Mohamed, H.H., Amal, A., Hussein, M.M., 2003. Influence of salinity on lipid per-oxidation, antioxidant enzymes, and electrophoretic patterns of proteins and isoenzymes in leaves of some onion cultivars. Asian J. Plant Sci. 2, 1220−1227. In: Nazar et al., 2011, Environ. Exp. Bot. 70, 80−87.

Eldardiry, E.I., Pibars, S.Kh., Abd El-Hady, M., 2012. Improving soil properties, maize yield components grown in sandy soil under irrigation treatments and humic acid application. Aust. J. Basic Appl. Sci. 6 (7), 587−593.

Eldardiry, E.I., Hellal, F., Mansour, H., Abd El-Hady, M., 2013. Assessment cultivated period and farmyard manure addition on some soil properties, nutrient content and wheat yield under sprinkler irrigation system. Agric. Sci. 4 (1), 14−22.

El-Enany, A.E., 1997. Glutathione metabolism in soybean callus-cultures as affected by salinity. Boil. Plant 39, 35−39. In: Nazar et al., 2011, Environ. Exp. Bot. 70, 80−87.

El-Hady, O.A., Abo-Sedera, S.A., 2006a. Conditioning effect of composts and acrylamide hydro-gels on a sandy calcareous soil. I- Physico-bio-chemical properties of the soil. Int. J. Agric. Biol. 8 (6), 876−884.

El-Hady, O.A., Abou-Sedera, S.A., 2006b. The conditioning effect of composts (natural) or/and acrylamide hydrogels (synthesized) on a sandy calcareous soil. II. Chemical and biological proper-ties of the soil. Egypt. J. Soil Sci. 46, 538−546.

El-Shakweer, M.H.A., El-Sayad, E.A., Ejes, M.S.A., 1998. Soil and plant analysis as a guide for interpretation of the improvement of efficiency of organic conditioners added to different soils in Egypt. Commun. Soil Sci. Plant Anal. 29, 2067−2088.

El-Shintinawy, F., El-Shourbagy, M.N., 2001. Alleviation of changes in protein metabolism in NaCl-stressed wheat seedlings by thiamine. Biol. Plant 44, 541−545.

Estrada, B., Aroca, R., Azcón-Aguilar, C., Barea, J.M., Ruiz-Lozano, J.M., 2013. Importance of native arbuscular mycorrhizal inoculation in the halophyte Asteriscus maritimus for successful establishment and growth under saline conditions. Plant Soil 370 (1−2), 175−185.

Fahad, S., Bano, A., 2012. Effect of salicylic acid on physiological and biochemical characteriza-tion of maize grown in saline area. Pak. J. Bot. 44 (4), 1433−1438.

FAO, 1996. Food and Agriculture Organization of the United Nations, Drainage of Irrigated Lands, Irrigation Water Management, Training Manual N 9, 79.

FAO, 2000. Food and Agriculture Organization of the United Nations, Global Network on Integrated Soil Management for Sustainable Use of Salt-affected. In: Lakhdar et al., 2009.

FAO, 2005. Management of Irrigation-Induced Salt-Affected Soils. Land and Water Development Div.; International Programme for Technology and Research in Irrigation and Drainage, Rome, Italy, pp. 1–4.

FAO, 2008. Land and plant nutrition management service. <http://www.fao.org/ag/agl/agll/spush/>.

FAO, 2015. Salt-affected soils: FAO Soils Portal. <http://www.fao.org/soils-portal/soil-management/management-of-some-problem-soils/salt-affected-soils/more-information-on-salt-affected-soils/en/>.

Fediuc, E., Lips, S.H., Erdei, L., 2005. O-Acetylserine (thiol) lyase activity in Phragmites and Typha plants under cadmium and NaCl stress conditions and the involvement of ABA in the stress response. J. Plant Physiol. 162, 865–872. In: Nazar et al., 2011, Environ. Exp. Bot. 70, 80–87.

Fei, G., YiJun, Z., LingYun, H., Da Cheng, H.E., GenFa, Z., 2008. Proteomic analysis of long-term salinity stress responsive proteins in Thellungiella halophila leaves. Chin. Sci. Bull. 53, 3530–3537, In: Nazar et al., 2011, Environ. Exp. Bot. 70, 80–87.

Frary, A., Gol, D., Keles, D., Okmen, B., Pınar, H., Sigva, H., et al., 2010. Salt tolerance in Solanum pennellii: antioxidant response and related QTL. BMC Plant Biol. 10, 58. In: Nazar et al., 2011, Environ. Exp. Bot. 70, 80–87.

Gaballah, M.S., Gomaa, A.M., 2004. Performance of faba bean varieties grown under salinity stress and biofertilized with yeast. J. Appl. Sci. 4, 93–99.

Gaffar, M.O., Ibrahim, Y.M., Wahab, D.A.A., 1992. Effect of farmyard manure and sand on the performance of sorghum and sodicity of soils. J. Indian Soc. Soil Sci. 40 (3), 540–543.

Ganieva, R.A., Allahverdiyev, S.R., Guseinova, N.B., Kavakli, H.I., Nafisi, S., 1998. Effect of salt stress and synthetic hormone polystimuline K on the photosynthetic activity of cotton (Gossypium hirsutum). Tr. J. Bot. 22, 217–221. In: Dong, 2012.

García, A.C., Izquierdo, F.G., González, O.L.H., Armas, M.M.D., López, R.H., Rebato, S.M., et al., 2013. Biotechnology of humified materials obtained from vermicomposts for sustainable agroecological purposes. Afr. J. Biotechnol. 12 (7), 625–634.

Gehad, A., 2003. Deteriorated Soils in Egypt: Management and Rehabilitation. Arab Republic of Egypt Ministry of Agriculture and Land Reclamation. Executive Authority for Land Impovement Projects (EALIP), pp. 1–36.

Gharaibeh, M.A., Eltaif, N.I., Shunnar, O.F., 2009. Leaching and reclamation of calcareous saline–sodic soil by moderately saline and moderate-SAR water using gypsum and calcium chloride. J. Plant Nutr. Soil Sci. 172 (5), 713–719.

Ghorbani, H., Babaei, A.A., 2008. The effects of natural zeolite on ions adsorption and reducing solution electrical conductivity I) Na and K solutions. In: International Meeting on Soil Fertility Land Management and Agroclimatology, Turkey, pp. 947–955.

Gil, R., Boscaiu, M., Lull, C., Bautista, I., Lidón, A., Vicente, O., 2013. Are soluble carbohydrates ecologically relevant for salt tolerance in halophytes? Funct. Plant Biol. 40, 805–818.

Gomaa, A.M., Gaballah, M.S., 2004. Changes in compatible solutes of some maize varieties grown in sandy soil and biofertilized with Rhodotorula glutinis under saline conditions. Zu finden. <http://www.bodenkunde2.uni-freiburg.de/eurosoil/abstracts/id105_Gaballah_full.pdf/> (Zitiert am 02.05.07.).

Gondim, F.A., Gomes-Filho, E., Lacerda, C.F., Prisco, J.T., Neto, A.D.A., Marques, E.C., 2010. Pretreatment with H_2O_2 in maize seeds: effects on germination and seedling acclimation to salt stress. Braz. J. Plant Physiol. 22 (2), 103–112.

González Núñez, L.M., Tóth, T., García, D., 2004. Integrated management for the sustainable use of salt-affected soils in Cuba Universidady Ciencia, 40 (20): 85–102.

Gorham, J., 1995. Mechanism of salt tolerance of halophytes. In: Choukr-Allah, R., Malcolm, C.V., Hamdy, A. (Eds.), Halophytes and Biosaline Agriculture. Marcel Dekker, New York, NY, pp. 207–233. ISBN-10: 0824796640.

Gossett, D.R., Millhollon, E.P., Lucas, M.C., 1994. Antioxidant response to NaCl stress in salt tolerant and salt-sensitive cultivars of cotton. Crop Sci. 34, 706–714. In: Nazar et al., 2011, Environ. Exp. Bot. 70, 80–87.

Grattan, S.R., Grieve, C.M., 1999. Mineral nutrient acquisition and response by plants grown in saline environments. In: Pessarakli, M. (Ed.), Handbook of Plant and Crop Stress. Marcel Dekker, New York, NY, pp. 203–229.

Gurmani, A.R., Bano, A., Ullah, N., Khan, H., Jahangir, M., Flowers, T.J., 2013. Exogenous abscisic acid (ABA) and silicon (Si) promote salinity tolerance by reducing sodium (Na^+) transport and bypass flow in rice (Oryza sativa indica). Aust. J. Crop Sci. 7 (9), 1219–1226.

Guru Devi, R., Pandiyarajan, V., Gurusaravanan, P., 2012. Alleviating effect of IAA on salt stressed Phaseolus mungo (L.) with reference to growth and biochemical characteristics. Recent Res. Sci. Technol. 4, 22–24.

Hajiboland, R., 2013. Role of arbuscular mycorrhiza in amelioration of salinity. Salt stress. In: Ahmad, P., Azooz, M.M., Prasad, M.N.V. (Eds.), Plants, pp. 301–354.

Hameed, A., Ahmed, M.Z., Gulzar, S., Khan, M.A., 2009. Effect of disinfectants in improving seed germination of Suaeda fruticosa under saline conditions. Pak. J. Bot. 41 (5), 2639–2644.

Hamid, N., Ali, S., Malik, K.A., Hafeez, F.Y., 2007. Diagnosis of nutritional constraints of Azolla spp. to enhance their growth under flooded conditions of salt affected soils. Pak. J. Bot. 39 (1), 161–167.

Hanafy Ahmed, A.H., Darwish, E., Hamoda, S.A.F., Alobaidy, M.G., 2013. Effect of putrescine and humic acid on growth, yield and chemical composition of cotton plants grown under saline soil conditions. Am. Eurasian J. Agric. Environ. Sci. 13 (4), 479–497.

Hare, P.D., Cress, W.A., 1997. Metabolic implications of stress-induced proline accumulation in plants. Plant Growth Regul. 21, 79–102.

Hassan, A.Z.A., Mahmoud, A.M., 2013. The combined effect of bentonite and natural zeolite on sandy soil properties and productivity of some crops. Topclass J. Agric. Res. 1 (3), 22–28.

Hatton, T.J., Nulsen, R.A., 1999. Towards achieving functional ecosystem mimicry with respect to water cycling in southern Australian agriculture. Agrofor. Syst. 45, 203–214.

Herschbach, C., Teuber, M., Eiblmeier, M., Ehlting, B., Ache, P., Polle, A., et al., 2010. Changes in sulphur metabolism of grey poplar (Populus × canescens) leaves during salt stress: a metabolic link to photorespiration. Tree Physiol. Available from: http://dx.doi.org/10.1093/treephys/tpq041In: Nazar et al., 2011, Environ. Exp. Bot. 70, 80–87.

Hillel, D., 2004. Introduction to Environmental Soil Physics. Elsevier Academic Press, Amsterdam, 494 pp.

Horneck, D.S., Ellsworth, J.W., Hopkins, B.G., Sullivan, D.M., Stevens, R.G., 2007. Managing Salt-Affected Soils for Crop Production. PNW 601-E. Oregon State University, University of Idaho, Washington State University.

Huchzermeyer, B., Flowers, T., 2013. Putting halophytes to work—genetics, biochemistry and physiology. Funct. Plant Biol. 40, 1–4.

Huez-López, M.A., Ulery, A.L., Samani, Z., Picchioni, G., Flynn, R.P., 2011. Response of chile pepper (Capsicum annuum) to salt stress and organic and inorganic nitrogen sources: ii. Nitrogen and water use efficiencies, and salt tolerance. Trop. Subtrop. Agroecosyst. 14, 757–763.

Hussein, M.M., Abou-Baker, N.H., 2014. Growth and mineral status of moringa plants as affected by silicate and salicylic acid under salt stress. Int. J. Plant Soil Sci. 3 (2), 163–177.

Hussein, M.M., Mehanna, H., Abou-Baker, N.H., 2012. Growth, photosynthetic pigments and mineral status of cotton plants as affected by salicylic acid and salt stress. J. Appl. Sci. Res. 8 (11), 5476–5484.

Ippolito, J.A., Tarkalson, D.D., Lehrsch, G.A., 2011. Zeolite soil application method affects inorganic nitrogen, moisture, and corn growth. Soil Sci. 176 (3), 136–142.

Irshad, A., Ahmad, I., Kim, S.B., 2013. Isolation, characterization and antimicrobial activity of halophilic bacteria in foreshore soils. Afr. J. Microbiol. Res. 7 (3), 164–173.

Jalaluddin, M., 1993. Effect of VAM fungus (Glomus intradices) on the growth of sorgum, maize, cotton and pennisetum under salt stress. Pak. J. Bot. 25, 215–218. In: Dong, 2012.

Javid, A., Yasin, M., Nabi, G., 2001. Effect of seed pre-treatments on germination and growth of cotton (Gossypium hirsutum L.) under saline conditions. Pak. J. Biol. Sci. 4, 1108–1110. In: Dong, 2012.

Jonáš, M., Salaš, P., Baltazár, T., 2012. Effect of exogenously application selected phytohormonal substances on the physiological and morphological indicators of philadelphus x hybrid in containers. Acta Univ. Agric. Silvic. Mendlianae Brun. 12 (8), 109–118.

Kahlaoui, B., Hachicha, M., Teixeira, J., Misle, E., Fidalgo, F., Hanchi, B., 2013. Response of two tomato cultivars to field-applied proline and salt stress. J. Stress Physiol. Biochem. 9 (3), 357–365.

Kant, A.C., Turan, M., 2011. Hydrogel substrate alleviates salt stress with increase antioxidant enzymes activity of bean (Phaseolus vulgaris L.) under salinity stress. Afr. J. Agric. Res. 6 (3), 715–724.

Kaya, C., Ashraf, M., Dikilitas, M., Tuna, A.L., 2013. Alleviation of salt stress-induced adverse effects on maize plants by exogenous application of indoleacetic acid (IAA) and inorganic nutrients—a field trial. Aust. J. Crop Sci. 7 (2), 249–254.

Kent, L.M., Läuchli, A., 1985. Germination and seedling growth of cotton: salinity-calcium interactions. Plant Cell Environ. 8, 155–159. In: Dong, 2012.

Khan, M.A., Gul, B., 2006. Halophyte seed germination. In: Khan, M.A., Weber, D.J. (Eds.), Ecophysiology of High Salt Tolerant Plants. Springer, Dordrecht, Netherlands, pp. 11–30. In: Hameed et al., 2009.

Khan, M.H., Singha Ksh, L.B., Panda, S.K., 2002. Changes in antioxidant levels in Oryza sativa L. roots subjected to NaCl-salinity stress. Acta Physiol. Plant 24, 145–148. In: Nazar et al., 2011, Environ. Exp. Bot. 70, 80–87.

Khan, N.A., Anjum, N.A., Nazar, R., Iqbal, N., 2009a. Increased activity of ATPsulfurylase, contents of cysteine and glutathione reduce high cadmium-induced oxidative stress in high photosynthetic potential mustard (Brassica juncea L.) cultivar. Russ. J. Plant Physiol. 56, 670–677. In: Nazar et al., 2011, Environ. Exp. Bot. 70, 80–87.

Khan, N.A., Nazar, R., Anjum, N.A., 2009b. Growth, photosynthesis and antioxidant metabolism in mustard (Brassica juncea L.) cultivars differing in ATP-sulfurylase activity under salinity stress. Sci. Hort. 122, 455–460. In: Nazar et al., 2011, Environ. Exp. Bot. 70, 80–87.

Khel, M., 2006. Saline soils of Iran with examples from the alluvial plain of Korbal, Zagros mountains, In: Soil and Desertification-Integrated Research for the Sustainable Management of Soils in Drylands Proceedings of the International Conference, Hamburg, Germany. In: Lakhdar et al., 2009.

Klocke, N.L., Currie, R.S., Aiken, R.M., 2009. Soil water evaporation and crop residues. TASABE 52 (1), 103–110.

Koprivova, A., Kopriva, S., 2008. Lessons from investigation of regulation of APS reductase by salt stress. Plant Signal. Behav. 8, 567–569. In: Nazar et al., 2011, Environ. Exp. Bot. 70, 80–87.

Koprivova, A., North, K.A., Kopriva, S., 2008. Complex signaling network in regulation of adenosine 5-phosphosulfate reductase by salt stress in *Arabidopsis* roots. Plant. Physiol. 146, 1408–1420. In: Nazar et al., 2011, Environ. Exp. Bot. 70, 80–87.

Kosová, K., Vítámvás, P., Urban, M.O., Prášil, I.T., 2013. Plant proteome responses to salinity stress—comparison of glycophytes and halophytes. Funct. Plant Biol. 40, 775–786.

Koupai, J.A., Eslamian, S.S., Kazemi, J.A., 2008. Enhancing the available water content in unsaturated soil zone using hydrogel, to improve plant growth indices. Ecohydrol. Hydrobiol. 8 (1), 67–75.

Koyro, H.W., Zörb, C., Debez, A., Huchzermeyer, B., 2013. The effect of hyperosmotic salinity on protein pattern and enzyme activities of halophytes. Funct. Plant Biol. 40, 787–804.

Lakhdar, A., Rabhi, M., Ghnaya, T., Montemurro, F., Jedidi, N., Abdelly, C., 2009. Effectiveness of compost use in salt-affected soil. J. Hazard. Mater. 171, 29–37.

Lechno, S., Zamski, E., Tel-Or, E., 1997. Salt stress-induced responses in cucumber plants. J. Plant Physiol. 150, 206–211. In: Nazar et al., 2011, Environ. Exp. Bot. 70, 80–87.

Li, Y., Song, X., Yang, X., 2008. Effects of seed soaking withglycinebetaine on the salt tolerance of cotton seedlings. Acta Agron. Sin. 34, 305–310. In: Dong, 2012.

Liang, Y.C., Yang, Y.F., Yang, C.G., Shen, Q.Q., Zhou, J.M., Yang, L.Z., 2003. Soil enzymatic activity and growth of rice and barley as influenced by organic matter in an anthropogenic soil. Geoderma 115 (1–2), 149–160.

Liang, Y., Sun, W., Zhu, Y.G., Christie, P., 2007. Mechanisms of silicon-mediated alleviation of abiotic stresses in higher plants: a review. Environ. Pollut. 147, 422–428.

Lopez-Berenguera, C., Carvajala, M., Garcea-Viguerab, C., Alcaraz, C.F., 2007. Nitrogen, phosphorus, and sulfur nutrition in Broccoli plants grown under salinity. J. Plant Nutr. 30, 1855–1870. In: Nazar et al., 2011, Environ. Exp. Bot. 70, 80–87.

Ma, J.F., Yamaji, N., 2006. Silicon uptake and accumulation in lower plants. Trends Plant Sci. 11 (8), 392–397.

MacCauley, A., Jones, C., 2005. Salinity and Sodicity Management, Soil and Water Management Module 2. Montana State University publication 4481-2. <http://landresources.montana.edu/swm/>.

Maksimovic, I., Ilin, Z., 2012. In: Lee, T.S. (Ed.), Effects of Salinity on Vegetable Growth and Nutrients Uptake, Irrigation Systems and Practices in Challenging Environments. InTech, p. 370. ISBN: 978-953-51-0420-9. <http://www.intechopen.com/books/irrigation-systems-and-practices-inchallenging- environments/effects-of-salinity-on-vegetable-growth-and-nutrients-uptake>.

Marschner, H., 1995. Mineral Nutrition of Higher Plants, second ed. Academic Press, New York, NY, In: Nazar et al., 2011, Environ. Exp. Bot. 70, 80–87.

Martínez-Ballesta, M.C., Moreno, D.A., Carvajal, M., 2013. The physiological importance of glucosinolates on plant response to abiotic stress in *Brassica*, review. Int. J. Mol. Sci. 14, 11607–11625.

Mashali, A., Suarez, D.L., Nabhan, H., Rabindra, R., 2005. Integrated Management for Sustainable Use of Salt-Affected Soils, FAO Soils Bulletin, Rome. In: Lakhdar et al., 2009.

Mittova, V., Theodoulou, F.L., Kiddle, G., Gomez, L., Volokita, M., Tal, M., et al., 2003. Coordinate induction of glutathione biosynthesis and glutathione metabolizing enzymes is correlated with salt tolerance in tomato. FEBS Lett. 554, 417–421. In: Nazar et al., 2011, Environ. Exp. Bot. 70, 80–87.

Mohamed, A.A., Eichler-Löbermann, B., Schnug, E., 2007. Response of crops to salinity under Egyptian conditions: a review. Landbauforsch. Voelkenrode 2 (57), 119–125.

Mohammad, M.J., Karam, N.S., Al-Lataifeh, N.K., 2004. Response of croton grown in a zeolite-containing substrate to different concentrationsof fertilizer solution. Commun. Soil. Sci. Plant Anal. 35 (15–16), 2283–2297.

Mohsen, A.A., Ebrahim, M.K.H., Ghoraba, W.F.S., 2013. Effect of salinity stress on *Vicia faba* productivity with respect to ascorbic acid treatment. Iranian J. Plant Physiol. 3 (3), 725–736.

Nazar, R., Iqbal, N., Masood, A., Syeed, S., Khan, N.A., 2011. Understanding the significance of sulfur in improving salinity tolerance in plants. Environ. Exp. Bot. 70, 80–87.

Noory, H., Liaghat, A., Chaichi, M.R., Parsinejad, M., 2009. Effects of water table management on soil salinity and alfalfa yield in a semi-arid climate. Irrig. Sci. 27, 401–407.

Olaiya, C.O., Gbadegesin, M.A., Nwauzoma, A.B., 2013. Bioregulators as tools for plant growth, development, defence and improvement. Afr. J. Biotechnol. 12 (32), 4987–4999.

Omer, E.A., Said-Al Ahl, H.A.H., El Gendy, A.G., Shaban, Kh.A., Hussein, M.S., 2013. Effect of amino acids application on production, volatile oil and chemical composition of chamomile cultivated in saline soil at Sinai. J. Appl. Sci. Res. 9 (4), 3006–3021.

Ondrasek, G., Rengel, Z., Veres, S., 2011. Soil salinisation and salt stress in crop production. In: Shanker A.K., Venkateswarlu B. (ur.). (Ed.), Abiotic Stress in Plants—Mechanisms and Adaptations, Rijeka, InTech, pp. 171–190. <https://bib.irb.hr/datoteka/559850. Ondrasek_Rengel_and_Veres.pdf>.

Parande, S., Zamani, G.R., Zahan, M.H.S., Ghader, M.G., 2013. Effects of silicon application on the yield and component of yield in the common bean (*Phaseolus vulgaris*) under salinity stress. Int. J. Agron. Plant Prod. 4 (7), 1574–1579.

Parida, K.A., Das, A.B., 2005. Salt tolerance and salinity effects on plants: a review. Ecotox. Environ. Saf. 60, 324–349.

Polat, E., Karaca, M., Demir, H., Onus, N., 2004. Use of natural zeolite (clinoptilolite) in agriculture. J. Fruit Ornamental Plant Res. 12 (Special ed.), 183–189.

Qadir, M., Oster, J.D., 2004. Crop and irrigation management strategies for saline–sodic soils and waters aimed at environmentally sustainable agriculture. Sci. Total Environ. 323 (1–3), 1–19.

Qadir, M., Noble, D.A., Schubert, S., Thomas, R.J., Arslan, A., 2006. Sodicity-induced land degradation and its sustainable management: problems and prospects. Land Degrad. Dev. 17 (6), 661–676.

Rameshkumar, S., Eswaran, K., 2013. Ecology, utilization and coastal management of salt tolerant plants (Halophytes and Mangroves) of mypad coastal regions, Andhra Pradesh, India. Int. J. Environ. Biol. 3 (1), 1–8.

Rengasamy, P., 2005. World salinisation with emphasis on Australia. Comp. Biochem. Phys. A 141, 337–348. In: Lakhdar et al., 2009.

Rengasamy, P., 2010. Soil processes affecting crop production in salt-affected soils. Funct. Plant Biol. 37, 613–620.

Rhoades, J.D., Loveday, J., 1990. Salinity in irrigated agriculture. In: Stewart, B.A., Nielsen, D.R. (Eds.), Irrigation of Agricultural Lands. Agronomy Monograph 30. American Society of Agronomy, Madison, WI, pp. 1089–1142.

Rhoades, J.D., Chanduvi, F., Lesch, S., 1999. Soil Salinity Assessment: Methods and Interpretations of Electrical Conductivity, Paper 57, FAO, Rome, Italy, pp. 1–150.

Rohanipoor, A., Norouzi, M., Moezzi, A., Hassibi, P., 2013. Effect of silicon on some physiological poroperties of maize (*Zea mays*) under salt stress. J. Biol. Environ. Sci. 7 (20), 71–79.

Romero, L.C., Domınguez-Solis, J.R., Gutierrez-Alcala, G., Gotor, C., 2001. Salt regulation of *O*-acetylserine(thiol)lyase in *Arabidopsis thaliana* and increased tolerance in yeast. Plant Physiol. Biochem. 39, 643–647, In: Nazar et al., 2011, Environ. Exp. Bot. 70, 80–87.

Roshandel, P., Flowers, T., 2009. The ionic effects of NaCl on physiology and gene expression in rice genotypes differing in salt tolerance. Plant Soil 315, 135–147. In: Nazar et al., 2011, Environ. Exp. Bot. 70, 80–87.

Ruiz, J.M., Blumwald, E., 2002. Salinity-induced glutathione synthesis in *Brassica napus*. Planta 214, 965–969.

Sahin, U., Eroglu, S., Sahin, F., 2011. Microbial application with gypsum increases the saturated hydraulic conductivity of saline–sodic soils. Appl. Soil Ecol. 48 (2), 247–250.

Sairam, R.K., Tyagi, A., 2004. Physiology and molecular biology of salinity stress tolerance in plants. Curr. Sci. 86, 407–721.

Savant, N.K., Datnoff, L.E., Snyder, G.H., 1997. Depletion of plant available silicon in soils: a possible cause of declining rice yields. Commun. Soil Sci. Plant Anal. 28 (13–14), 1245–1252.

Scherer, H.W., 2008. Impact of sulphur on N_2 fixation of legumes. In: Khan, N.A., Singh, S., Umar, S. (Eds.), Sulfur Assimilation and Abiotic Stresses in Plants. Springer-Verlag, New York, NY, pp. 43–54. In: Nazar et al., 2011, Environ. Exp. Bot. 70, 80–87.

Schnug, E., Haneklaus, S., Murphy, D., 1993. Impact of sulfur fertilization on fertilizer nitrogen efficiency. Sulfur Agric. 17, 8–12. In: Nazar et al., 2011, Environ. Exp. Bot. 70, 80–87.

Shaharoona, B., Arshad, M., Zahir, Z.A., 2006. Effect of plant growth promoting rhizobacteria containing ACC-deaminase on maize (*Zea mays* L.) growth under axenic conditions and on nodulation in mung bean (*Vigna radiata* L.). Lett. Appl. Microbiol. 42, 155–159.

Shalata, A., Tal, M., 1998. The effect of salt stress on lipid peroxidation and antioxidants in the leaf of the cultivated tomato and its wild salt-tolerant relative *Lycopersicon pennellii*. Physiol. Plant 104, 169–174. In: Nazar et al., 2011, Environ. Exp. Bot. 70, 80–87.

Shalata, A., Neumann, P.M., 2001. Exogenous ascorbic acid (vitamin C) increases resistance to salt stress and reduces lipid peroxidation. J. Exp. Bot. 52, 2207–2211.

Shanker, A.K., Venkateswarlu, B., 2011. Abiotic Stress in Plants–Mechanisms and Adaptations. Janeza Trdine, Rijeka, Croatia, pp. 1–428.

Shao, H.B., Chu, L.Y., Jaleel, C.A., 2008. Water deficit stress induced anatomical changes in higher plants. C. R. Biol. 331 (3), 215–225.

Sharif, F., Khan, A.U., 2009. Alleviation of salinity tolerance by fertilization in four thorn forest species for the reclamation of salt-affected sites. Pak. J. Bot. 41 (6), 2901–2915.

Shinde, S.K., Shinde, B.P., Patale, S.W., 2013. The alleviation of salt stress by the activity of AM fungi in growth and productivity of onion (*Allium cepa* L.) plant. Int. J. Life Sci. Pharma Res. 3 (1), 11–15.

Silva, C.M.M.de S., Fay, E.F., 2012. Effect of Salinity on Soil Microorganisms. In: Hernandez Soriano, M.C. (Ed.), Soil Health and Land Use Management. Rijeka: InTech, pp. 177–198. <http://ainfo.cnptia.embrapa.br/digital/bitstream/item/73177/1/2012CL03.pdf>.

Smith, T.E., Grattan, S.R., Grieve, C.M., Poss, J.A., Suarez, D.L., 2010. Salinity's influence on boron toxicity in broccoli: I. Impacts on yield, biomass distribution, and water use. Agric. Water Manage. 97 (6), 777–782.

Stamford, N.P., Silva, A.J.N., Freitas, A.D.S., Araujo Filho, J.T., 2002. Effect of sulphur inoculated with *Thiobacillus* on soil salinity and growth of tropical tree legumes. Bioresour. Technol. 81, 53–59.

Stark, C., 1991. Osmotic adjustment and growth of salt-stressed cotton as improved by a bioregulator. J. Agron. Crop Sci. 167, 326–334. In: Dong, 2012.

Suzuki, N., Rivero, R.M., Shulaev, V., Blumwald, E., Mittler, R., 2014. Abiotic and biotic stress combinations. New Phytol. 203, 32–43.

Tawfik, M.M., Thalooth, A.T., Zaki, N.M., Hassanein, M.S., Bahr, A.A., Ahmed, A.G., 2013. Sustainable production of *Kochia indica* grown in saline habitat. J. Environ. Treat. Tech. 1 (1), 56–61.

Tejada, M., Garcia, C., Gonzalez, J.L., Hernandez, M.T., 2006. Use of organic amendment as a strategy for saline soil remediation: influence on the physical, chemical and biological properties of soil. Soil Biol. Biochem. 38 (6), 1413–1421.

Torabian, A.R., 2011. Effect of salicylic acid on germination and growth of alfalfa (*Medicago sativa* L.) seedlings under water potential loss at salinity stress. Plant Ecophysiol. 2, 151–155.

Tòth, G., Montanarella, L., Rusco, E., 2008. Updated map of salt affected soils in the european union. Threats to Soil Quality in Europe. European Communities, Luxembourg, pp. 61–74.

Tripathi, S., Kumari, S., Chakraborty, A., Gupta, A., Chakrabarti, K., Bandyapadhyay, B.K., 2006. Microbial biomass and its activities in salt-affected soils. Biol. Fertil. Soils 42 (3), 273–277.

Tuna, A.L., Kaya, C., Higgs, D., Murillo-Amador, B., Aydemir, S., Girgin, A.R., 2008. Silicon improves salinity tolerance in wheat plants. Environ. Exp. Bot. 62, 10–16.

Varallyay, G., 1992. Soil data base for sustainable land use—Hungarian case study. In: Proceedings of the International Symposium on Soil Resilience and Sustainable Land Use. Budapest, September 28–October 2. In: Lakhdar et al., 2009.

Virta, R.L., 2000. Zeolites. U.S. Geological Survey minerals Information, In: U.S. Geological Survey Publication Zeolite. Ch. in United States Mineral Resources, U.S. Geological Survey Professional Paper 820. p. 1–3. < http://minerals.usgs.gov/minerals/pubs/commodity/zeolites/zeomyb96.pdf >.

Waseem, R., Preeti, R., Suchit, A.J., Pramod, W.R., 2012. *Azolla-Anabaena* association and its significance in supportable agriculture. J. Biol. Chem. 40 (1), 1–6.

Waskom, R.M., Bauder, T.A., Davis, J.G., Andales, A.A., 2014. Diagnosing saline and sodic soil problems. Colorado State University Extension Fact Sheet # 0.521. <http://www.ext.colostate.edu/pubs/crops/00521.html>.

Watanabe, K., Tanaka, T., Hotta, Y., Kuramochi, H., Takeuchi, Y., 2000. Improving salt tolerance of cotton seedlings with 5-aminolevulinic acid. Plant Growth Regul. 32, 99–103. In: Dong, 2012.

Wichelns, D., 1999. An economic model of waterlogging and salinization in arid regions. Ecol. Econ. 30, 475–491. In: Lakhdar et al., 2009.

Williams, J., 2010. Salt in our cities—counteracting the silent flood. In: Urban Salt Conference, Parramatta, NSW. In: Ondrasek et al., 2011.

Wu, Q.S., Zou, Y.N., 2009. Adaptive responses of birch-leaved pear (*Pyrus betulaefolia*) seedlings to salinity stress. Not. Bot. Horti Agrobot. Cluj-Napoca 37, 133–138. In: Nazar et al., 2011, Environ. Exp. Bot. 70, 80–87.

Xianzhao, L., Chunzhi, W., Qing, S., 2013. Screening for salt tolerance in eight halophyte species from Yellow River Delta at the two initial growth stages. ISRN Agronomy, Hindawi Publishing Corporation, Volume 2013, Article ID 592820. 8 pages. Available from: http://dx.doi.org/10.1155/2013/592820.

Xie, Z., Duan, L., Tian, X., Wang, B., Eneji, A.E., Li, Z., 2008. Coronatine alleviates salinity stress in cotton by improving the antioxidative defense system and radical-scavenging activity. J. Plant Physiol. 165, 375–384, In: Dong, 2012.

Yamada, M., Uehira, M., Hun, L.S., Asahara, K., Endo, T., Eneji, A.E., et al., 2002. Ameliorative effect of K-type and Ca-type artificial zeolites on the growth of beets in saline and sodic soils. Soil Sci. Plant Nutr. 48 (5), 651–658.

Yan-min, Y., Xiao-jing, L., Wei-qiang, L., Cun-zhen, L., 2006. Effect of different mulch materials on winter wheat production in desalinized soil in Heilonggang region of North China. J. Zhejiang Univ. Sci. B (Biomed. Biotechnol.) 7 (11), 858–867.

Yao, K., Wu, Z., Zheng, Y., Kaleem, I., Li, C., 2010. Growth promotion and protection against salt stress by *Pseudomonas putida* Rs-198 on cotton. Eur. J. Soil Biol. 46, 49–54. In: Dong, 2012.

Yasuda, H., Takuma, K., Fukuda, T., Araki, Y., Suzuka, J., Fukushima, Y., 1998. Effect of zeolite on water and salt control in soil. Tottori Daigaku Nogakubu Kenkyu Hokoku Journal, Bulletin of the Institute of Tropical Agriculture, Kyushu University 51, 35–42. <http://agris.fao.org/agris-search/search.do?recordID=JP1999002954/>.

Yuan, B.C., Li, Z.Z., Liu, H., Gao, M., Zhang, Y.Y., 2007. Microbial biomass and activity in salt affected soils under arid conditions. Appl. Soil Ecol. 35 (2), 319–328.

Yue, H.T., Mo, W.P., Li, C., Zheng, Y.Y., Li, H., 2007. The salt stress relief and growth promotion effect of Rs-5 on cotton. Plant Soil 297, 139–145. In: Dong, 2012.

Yuvaniyama, A., 2001. Managing problem soils in northeast Thailand. In: Kam, S.P., Hoanh, C. T., Trebuil, G., Hardy, B. (Eds.), Natural Resource Management Issues in the Korat Basin of Northeast Thailand: An Overview, vol. 7. IRRI Limited Proceedings, pp. 147–156. In: Lakhdar et al., 2009.

Zaka, M.A., Obaid-Ur-Rehman, R.H.U., Khan, A.A., 2005. Integrated approach for reclamation of salt affected soils. J. Agric. Soc. Sci. 1 (2), 94–97. <http://www.Ijabjass.org/>.

Zhu, J., Fu, X., Koo, Y.D., Zhu, J.K., Jenney Jr, F.E., Adams, M.W.W., et al., 2007. An enhancer mutant of *Arabidopsis* salt overly sensitive 3 mediates both ion homeostasis and the oxidative stress response. Mol. Cell. Biol. 27, 5214–5224.

Zhu, Z.J., Fan, H.F., He, Y., 2011. Roles of silicon-mediated alleviation of salt stress in higher plants: a review. In: International Conference on Silicon in Agriculture, Proceedings of the 5th, September, Beijing, China, p. 223.